Research Reports ESPRIT

Project 2054 · UCOL · Volume 1

Edited in cooperation with
the Commission of the European Communities

S. Forcesi (Ed.)

Ultra-wideband
Coherent Optical LANs

Springer-Verlag

Berlin Heidelberg New York
London Paris Tokyo
Hong Kong Barcelona
Budapest

Editor

S. Forcesi
Alcatel Italia S.p.A.
Via Abruzzi 25, I-00187 Roma, Italy

ESPRIT Project 2054 "Ultra-wideband Coherent Optical LAN (UCOL-2)" is part of the technological area "Advanced Business and Home Systems—Peripherals" of the ESPRIT Programme (European Specific Programme for Research and Development in Information Technology) supported by the European Communities.

This project targets the enhanced performance and flexibility that can be built into a multichannel, ultra-wideband network using coherent optical techniques. The system concept is based on multichannel optical transmission with heterodyne detection, where all optical channels are synchronised to an optical reference comb. Work has evolved through a series of phases of experimental development and has led to the production of practical integrated optical components and subsystems, which are described in this book.

Partners in the project are:
Alcatel FACE S.p.A. (Italy) · SEL Alcatel (Germany) · Koninklijke PTT Nederland (The Netherlands) · University of Southampton (UK) · Daimler-Benz (Germany) · Alcatel Alsthom Recherche (France) · GEC Hirst (UK) · INESC (Portugal) · Telettra Espanola S.A. (Spain) · COSI (Italy) · IDATE (France)

CR Subject Classification (1991): B.4, C.2

ISBN-13: 978-3-540-56885-8   e-ISBN-13: 978-3-642-45733-3
DOI: 10.1007/978-3-642-45733-3

Publication No. EUR 14972 EN of the
Commission of the European Communities, Dissemination of Scientific and Technical Knowhow Unit,
Directorate General Information Technologies and Industries, and Telecommunications,
Luxembourg

Typesetting: Camera - ready by authors
45/3140 – 543210 – Printed on acid-free paper

Preface

The content of this book is the result of the work and the experiences of an interdisciplinary and strictly European group of researchers who have attempted to give birth to a new fibre communication network concept by exploiting the potential benefits of optical coherent transmission.

The run towards this ambitious goal started in 1985 on the basis of the ESPRIT program incentives by an "ad hoc" consortium of industrial partners and research institutions. The first three years were dedicated to a feasibility study carried out by a reduced number of "pioneers". In 1989 the team was extended to eleven partners. There was already clear evidence of the increasing interest in Europe in exploring the actual limits of optical fibre coherent systems; the project had innovative and very advanced features that were gradually refined by incorporating the latest technological developments to which it has directly contributed.

The main objective of the project targeted the development of the necessary building blocks to show experimentally the performance and the flexibility built into the proposed concept of multi-channel ultra-wideband network. The system concept developed within this project associates optical frequency division multiplexing with a suitable network architecture and management techniques to allow very high flexibility and efficiency in handling simultaneous transmission over the network (on each optical frequency) of multiple virtual channels each operating with a wide range of information rates ranging from a fraction of a Mb/s up to a maximum value in excess of 160 Mb/s. Specialised applications requiring individually more than the capacity provided by a single optical channel were to be dealt with by parallel transmission over more than one optical channel.

Another important objective of this system concept was to make the network open to any future evolution through its capability to self-reconfigure without physical changes.

The book consists of a collection of technical reports and therefore is intended as a support for those professionals who wish to take advantage of the design experience actually gained in the context of an industrial enviroment. It has no aspiration to be a traditional academic text. In fact, even in the first pages, the reader is faced with a table reporting a series of expected performances of the studied network (named UCOL) that assumes a background knowledge of the main concepts of digital communications.

Subsequently, the rationale for the existence of a network such as UCOL and its envisaged position in the telecommunications world are given.

Chapter 3 explains, with some details, the architecture of the network and of its stations, as well as the original protocol developed in the context of OSI structure for the network management. The need to avoid non-open solutions led to the adoption of ATM techniques to ensure both a network architecture independent of the specific services and easy interworking with public and local networks that will be based on ATM, thereby minimizing the gateway functions.

In Chapter 4, attention is addressed to the criteria used to determine system parameters also as a consequence of the defined optical hardware.

Chapter 5 summarizes the potential options for an upgrading of the proposed system by optical amplification.

Finally, Chapters 6 and 7 report experimental results and technical information concerning, respectively, devices and subsystems developed so far, and the realized link between transmitter and receiver, both locked to a common reference by means of AFC techniques.

A very complete and up-to-date bibliography, including also the authors' publications, concludes the book.

Spring 1993   S. Forcesi

Contents

1   Introduction   1

2   UCOL Baseline   3

3   UCOL Network Architecture   9
3.1   UCOL Station Architecture   17

4   UCOL Optical Architecture   31
4.1   Introduction   31
4.2   Optical Receiver   33
4.3   Forward Error-correcting Coding   34
4.4   Channels Allocation Plan   35
4.5   Reference Carrier Generation   37
4.6   Carrier Generation for the Individual Channels   37

5   Network Extension   39

6   Experimental Results: Developed Components   43
6.1   Multi-line Master Source: Mode-Locked Laser   43
6.2 Er^{3+}-doped Tunable Narrow-line Laser   46
6.3   Active Phase Modulator and Switch   46
6.4   Polarisation Diversity DPSK Receiver   48
6.4.1   Optical Hybrid   48
6.4.2   Low-Noise Preamplifier   50
6.4.3   DPSK Demodulator   52
6.5   Erbium-Doped Fibre Amplifier   57
6.6   Tunable Optical Filter   60
6.7   Traffic Generator Detector   62
6.7.1   TGDS Architecture   62
6.7.2   Hardware Development   64
6.7.3   Software Developments   65
6.7.4   Results   68

7	**Experimental Results: Implementation of a Link between Transmitter and Receiver**	**71**
7.1	Demonstrator Set-up	71
7.1.1	Transmission Link	73
7.1.2	Lasers	74
7.1.3	Phase Modulators	74
7.1.4	Integration of Hybrid, Front-end and Amplifier	74
7.1.5	Comb Generator	75
7.1.6	Polarization Scrambler	76
7.1.7	Automatic Frequency Control	76
7.1.8	Optical Network	77
7.1.9	BER Measurement Set-up	78
7.2	Experiments	78
7.2.1	Frequency Locking	78
7.2.2	DPSK Transmission Link	79
7.3	Results and Conclusions	84
7.3.1	Passive Phase Modulation	84
7.3.2	Active Phase Modulation	85
7.3.3	Comparison Between Active and Passive Phase Modulation	86
7.3.4	Conclusions	86
8	**UCOL Bibliography**	**87**
8.1	Technical Reports	87
8.2	Publications	90
	References	**94**
	UCOL Consortium	**95**
	Authors' Addresses	**96**

List of Abbreviations

AAL	ATM Adaptation Layer
AC_ME	Access Control ME
AC_PMD	Access Control PMD
ACI	Adjacent Channel Interference
ACU	Access Control Unit
AFC	Automatic Frequency Control
AGC	Automatic Gain Control
AM	Amplitude Modulation
ANSI	American National Standard Institute
ASE	Amplified Spontaneous Emission
ASK	Amplitude Shift Keying
ATM	Asynchronous Transfer Mode
AWGN	Additive White Gaussian Noise

B-ISDN	Broadband ISDN
BCH	Bose-Chaudhuri-Hocquenghem
BER	Bit Error Rate
BR	Bit Rate
BW	BandWidth

CBO	Continuous Bit-stream Oriented
CCI	Co-Channel Interference
C/I	Carrier-to-Interference ratio
CL	Connection-Less
CMC	Coherent Multi-Channel
CO	Connection Oriented
CRC	Clock Recovery Circuit

DA	Destination Address
DBR	Distributed Bragg Reflector
DD	Direct Detection
DFB	Distributed Feed-Back
DPSK	Differential Phase Shift Keying
DQAC	Distributed Queue Access Control

EDFA	Erbium-Doped Fibre Amplifier
EEC	European Economic Community
ESA	Excited State Absorption

FBO	Fixed Bit-stream Oriented
FBT	Fused Biconic Taper
FDDI	Fibre Distributed Data Interface
FDM	Frequency Division Multiplexing
FEC	Forward Error-correction Coding
FET	Field-Effect Transistor
FSK	Frequency Shift Keying
FSR	Free Spectral Range
FTTC	Fibre-To-The-Curb
FTTH	Fibre-To-The-Home

GAU	Group Access Unit

HD_WDM	High Density WDM
HDTV	High Definition TV
HiBi	High Birefringent
HSC	High Speed Channel

IBCN	Integrated Broadband Comm. Network
IEEE	Institute of Electrical and Electronic Engineering
IF	Intermediate Frequency
IM	Intensity Modulation
ISDN	Integrated Service Digital Network
ISI	Inter-Symbol Interference
ISO	International Standard Organization

L_ME	Layer ME
LAN	Local Area Network
LLC	Logical Link Control

LO	Local Oscillator
MAC	Medium Access Control
MAN	Metropolitan Area Network
ME	Management Entity
MIB	Management Information Base
MID	Message IDentifier
MLL	Mode-Locked Laser
MMI	Man-to-Machine Interface
N_ME	Network ME
NA	Numerical Aperture
NF	Noise Figure
NI	Network Interface
OPLL	Optical Phase Locked Loop
OSI	Open System Interconnection
PB	Peripheral Bus
PCB	Printed Circuit Board
PDU	Protocol Data Unit
PM	Phase Modulation
PMD	Physical Medium Dependent
PMF	Polarisation Maintaining Fibre
PSK	Phase Shift Keying
PU	Processing Unit
QOS	Quality Of Service
QS	Queue Status
RF	Radio Frequency
RGB	Reference Generation Block
RIN	Relative Intensity Noise
RQBW	ReQuested BandWidth
RSBW	ReServed BandWidth
RX	Receiver
S_M_AAL	Signalling & Management AAL
S_ME	Station ME
S_MIB	Station MIB
S_PMD	Station PMD
SA	Service Address
SCC	Switch Connection Control
SCF	Signalling Convergence Function
SDH	Synchronous Digital Hierarchy
SDU	Service Data Unit

SE	Session Execution
SE_ME	Station Element ME
SIA	Signalling Information Access
SLA	Semiconductor Lasers Amplifier
SM	Session Management
SMF	Single-Mode Fibre
SMU	Station Management Unit
S/N	Signal-to-Noise ratio
SOP	State Of Polarization
SVN	Service Virtual Network
TA	Terminal Adapter
TDM	Time Division Multiplexing
TDMA	Time Division Multiple Access
TE	Terminal Equipment
TE/TM	Transv. Electric / Transv. Magnetic
TGDS	Traffic Generator/Detector System
TNL	Tunable Narrow-line Laser
TX	Transmitter
U_AAL	User AAL
UAG	User Access Group
UAU	User Access Unit
UCOL	Ultra-wideband Coherent Optical LAN
UI	User Interface
UIA	User Information Access
UT	UCOL Termination
UT_ME	UCOL Termination ME
VBR	Variable Bit Rate
VCI	Virtual Channel Identifier
VPI	Virtual Path Identifier
WAN	Wide Area Network
WDM	Wavelength Division Multiplexing
WS	WorkStation

1 Introduction

The objective of this project has targeted the provision of cost-effective local area wideband communication capacity to meet primarily industrial and commercial requirements. The project addressed both the technological and systems aspects of a flexible and efficient communication network.

This project followed activity commenced in February 1985, with the financial support from the EEC ESPRIT program, which explored the feasibility of employing optical FDM in a star-based topology. Previous work of the partnership in this field selected the network architecture, access protocols and management methods. Optimization of the overall system in terms of cost, geographical extension, flexibility and future user requirements has been the fundamental element of work. The longer term aim of the work carried out in this project has been the realization of a number of components for a prototype demonstrator capable of supporting wide-band, multi-service operation.

This work stemmed from the now recognized fact that any effective integrated office communication system which connects high traffic density equipments such as graphic work-stations, computers and video terminals must have a bandwidth substantially greater than that offered by current LANs. It is apparent that the explosive growth in the use of inter-connected data processing equipment, foreseeable within a 5 to 10 years period at most, will demand network capacity in excess of several Gb/s.

This project has been set out to demonstrate use of coherent optical techniques in local networks. Such techniques afford a system capacity of the order of tens of Gb/s and offer a cost effective use of the technology because the overall capacity can be made available to a large number of users in a very flexible way.

At the physical layer, the system concept is based on multichannel optical transmission with heterodyne DPSK detection where all optical channels are stabilised using a distributed referencing concept. Work has evolved through a series of phases of experimental development, leading to the fabrication of optical components and subsystems.

Exploitation of the results of the project will take place with different time scales. In the earlier phase components and subsystems developed within the project will lead to prototypes likely to be used in a number of potential products differing in features and applications. In a later phase also the system concepts developed within the project will find their way towards commercial exploitation together with the technology developed to support it.

Details on the expected performance of the system concept are given in Table 1.1.

Table 1.1 - Expected performance of UCOL

Number of Stations	max.	64
Number of GAUs per Station	max.	15
Number of UAUs per GAU	max.	32
Number of UAUs per Station	max.	480
Number of ACUs per Station	max.	15
Station ATM Switch configuration	min.	8×8
	max.	32×32
Number of NIs per Station	max.	15
Total number of NIs	max.	960
Number of Optical Channels		20
Channel net bit-rate		163.584 Mb/s
Channel gross bit-rate		300 Mb/s
Station net capacity	max.	~ 2.5 Gb/s
Network net capacity	max.	> 3.2 Gb/s
CBO services		64 kb/s + 163.584 Mb/s
Bursty services		up to 163.584 Mb/s
Connection set-up time (connection-oriented traffic)		20 ms
Network access time (connectionless traffic)		2 ms
End-to-end network delay time (isochronous traffic)		< 2 ms
Quality of service:	ATM cell lost/Yr	0.8
	ATM cell with errors/Yr	15
Multi-channel scheme		FDM
Channel access type		TDMA
Topology		Central STAR
Star size	max. diameter 20 km	1024×1024
	max. diameter 50 km	256×256
Modulation scheme		DPSK
System gain	(at BER=10^{-6})	47.9 dB

2 UCOL Baseline

A. Fioretti

The technical context of work on the project can be provided by a critical discussion of the main alternatives available to meet the objectives of the project, i.e. the commercial provision of integrated support for practical narrowband and broadband services.

In the past few years the traffic exchanged inside confined areas has increased rapidly; today, capacities of 100 Mb/s are becoming common. The FDDI standard, which is already marketed, provides an example of such networks. The European market for LANs is growing at a rate around 40% per year. Forecasts for the future confirm this trend and it is to be expected that in a few years the market will be mature to accept networks with capacities of Gb/s.

This can be evinced from the following considerations:

a) Up to now networks have been basically used to connect terminals with hosts, that is, in practice to share one or more resources between several users distributed in a predefined area. Therefore LANs have not been used to achieve resource distribution but only the distribution of the services provided by a common resource. The distribution of resources leads to a cost effective solution avoiding resource redundancy and at the same time provides increased system reliability and a more elastic and powerful use of the resources. Nevertheless, resource distribution implies the direct connection between different machines; as a consequence the rate of exchanged information increases considerably. From this point of view local networks can be seen as a distributed computer's serial bus. Considering that today 32 bits microprocessors are capable of moving data at a rate up to 200 Mbytes/s it follows immediately that in a distributed system the communication subsystem must provide a bandwidth of Gb/s in order to meet the requirements of a fully distributed system.

b) Up to now due to the relatively low capacity of conventional networks, services related to real time video transmission have not been developed. With the new generation of Gb/s networks it is possible to start stimulating new services and to prepare the market to exploit them.

c) A very special application generating the need for high speed communication systems stems from the continuous growth in high-end computing power, notably in "supercomputers" or "near-supercomputers" based environments. The specific application that appears most demanding in terms of individual channel bit rate is "visualization" [1], that is the capacity of visualizing in real time a model (program) executing on a supercomputer. The parameters that give in quantitative terms the transmission capacity needed to exchange moving images (for example between the computer and a workstation) can be obtained considering a video transmission with a rate of 24 frames/sec. If each frame has a resolution of 1024×1024 pixels and each pixel is coded with 8 bits per color a point-to-point connection with a bandwidth of roughly 600 Mb/s is required. The lack of open architectures in the solutions currently adopted to meet these requirements constitutes a serious hurdle to their development into networks capable of offering a multi-service environment. The only emerging standard that aims to fulfill these needs is the *High Speed Channel* (HSC), a point-to-point link proposed by an ANSI Committee, which describes an interface operating at peak data rate of 800 or 1600 Mb/s over copper cabling at distances up to 25 meters. The transfer rate of 800 (1600) Mb/s is obtained with a 32 (64) bit-wide parallel bus running at the frequency of 25 MHz. It has however to be stressed that the requirement for a very high speed channel in the specific application mentioned above is likely to represent a rather small fraction of the potential market for high capacity networks as the application is restricted to a small number of users belonging to a limited scientific community.

d) Another application representative of a more promising and wider class of applications where emphasis is on total network capacity rather than individual channel capacity is the automated production of daily newspapers relying on a system designed around laser film printers, high resolution scanners, workstations and fileservers adequately networked. The editorial work is performed at workstations, while everything is stored in a fileserver with the problems of coordinating the various operators that contribute to the same article or advertisement and providing the most updated version through the several editing cycles. The Houston Chronicle, the first daily newspaper that has undertaken to achieve the complete computerized production, has identified in the next three-four years time frame the necessity of a backbone system able to interconnect 800 workstations and 20 fileservers with an overall bandwidth requirements of 1+2 Gb/s [2]. Therefore networks with a capacity in the order of Gb/s, able to connect a large number of users with different throughput needs and easily expandable to cope with the rather-unpredictable rate of market penetration of supercomputers will become common in the next decade.

The comparison of the technology alternatives that will support these networks deals with the following points:

- single channel versus multichannel
- High density WDM or WDM versus FDM
- geographic extension
- capacity and flexibility
- existing LAN / ISDN / future B-ISDN connectivity

Concerning the first point the advantages offered by the multichannel approach can be summarized as follows:

a) the attainable bandwidth is wider and it may be extended by increasing the number of frequencies, with small impact on pre-existing hardware and software.
b) as the optical technology becomes more mature the performance (both in terms of cost and actual performance) of a network based on the second approach will improve considerably.
c) the use of multi-channelling allows easy handling of wideband services by means of dedicated channels. Such application requires the capability to increase the bandwidth allocated to a single service by grouping the capacity of multiple channels.

With respect to the second point it should be stressed that the tunability of optical frequencies makes the network flexible and opened to future evolution because of its capability to self-reconfigure: in fact by suitably reallocating frequencies, it is possible to reconfigure "virtually" the network according to new requirements without any physical change. This last aspect is substantial because it shows one of the main advantages of the multichannel system over the competing technique: Wavelength Division Multiplexing, which although in principle could offer an alternative solution, in practice is severely limited by the rigid channeling scheme and the wide channel separation.

The technical literature describes a number of experiments on optical multi-channel systems. The methods adopted can be divided in three major groups:

- *Wavelength Division Multiplexed* (WDM) systems, in which a limited number of optical channels, spaced by 1 or more nanometers and multiplexed by a central star, are demultiplexed by WDM optical filters and directly detected;
- *High Density Wavelength Division Multiplexed* (HDWDM) systems, in which a great number of optical channels, with a frequency spacing in the order of ten times the bit-rate, can be separately accessed by means of tunable optical filters followed by direct detection receivers;
- *Frequency Division Multiplexed* (FDM) systems, which can be distinguished from HDWDM systems because of the demodulation scheme, which in this last case adopts heterodyne detection by means of tunable LO lasers.

WDM systems have the advantage, over HDWDM and FDM, of their relatively simple implementation, as they need neither complicated receiver schemes nor narrow-linewidth and highly stabilized lasers. Major drawbacks are:

- low receiver sensitivity;
- low number of available channels.

The first point contribute to lower the available system gain; thus although WDM systems are suitable for high capacity interoffice networks they cannot be used equally well for applications requiring both high number of subscribers and substantial geographical extension. Furthermore, to obtain high capacity networks with just a few available optical channels, high bit-rate transmission is required which in turn implies the need of high speed electronics at the receiver end; therefore, in order to obtain the same flexibility offered by FDM (and HDWDM) systems, a number of such high speed receivers (one for each optical channel) is required: this is unlikely to make this solution cost effective.

HDWDM systems are based on tunable optical filters, which do not allow the same narrowband filtering provided by the RF filters of coherent receivers but, for high bit-rate, theorically approach the same number of available optical channels of FDM systems, yet at the expense of reduced receiver sensitivity.

FDM seems therefore to represent the best choice when high capacity multipoint to multipoint applications are considered together with high geographical extension, high number of connected users and flexibility.

The advantages of coherent techniques in comparison with the complexity of the technical solution and the associated costs are still debated. Direct detection is by now based on well known components and technical solutions while devices for coherent optics are still in the development phase and often interim solutions are necessary to overcome performance gap. Also costs of optical devices needed for coherent transmission are high due to the low production and low yield.

It is perceived, however, that this situation is temporary and that the cost and complexity gap between direct and coherent technique will become rather limited. For instance it is expected that monolithic lasers would replace in the near future external cavity lasers simplifying both the transmitter and the receiver. Besides, the development of technology and the diffusion of coherent technique for point-to-point applications are likely to make the costs associated to this solution comparable to direct detection ones.

Networks capacity in the Gb/s range will appear in the immediate future; extrapolating this trend, capabilities of several Gb/s will be required before year 2000. Such a performance could only be fulfilled by a multichannel network equipped with a management able to supervise both the optical layer and the higher layers of network architecture.

Flexibility is certainly going to be the key requirement on both network architecture and management scheme; flexibility refers to the network capability to exploit efficiently the available bandwidth and adequately supporting any quality of service. This plays a decisive role in the degree of acceptance of the network by the wide market of potential users. This concept has been considered a primary objective during the evolution stages of the UCOL project, by pursuing service independent solutions.

Besides the advantages coming from the adoption of FDM techniques, which offer a manageable segmentation of the total bandwidth while making it possible, if required, to provide a single service with a bandwidth higher than that of the single channel (by assigning several channels in parallel), further levels of flexibility have been added:

- firstly, by time multiplexing each channel among different users;
- secondly by using a common access protocol for both isochronous and non-isochronous traffic;

- and finally by providing a prioritized mechanism to meet different service requirements, particularly in terms of medium access delay.

These solutions lead to the capability of supporting CBO traffic ranging from 64 kb/s to 163.5 Mb/s and bursty traffic up to 163.5 Mb/s on each optical channel.

3 UCOL Network Architecture

A. Bianchi, D. Capolupo, F. del Castello and A. Fantini

Previous work on UCOL has been carried out following three parallel lines:

1) Efficient exploitation of the capacity of each single optical channel by sharing its available bandwidth between several users by means of TDMA.
2) Management of the available frequencies to guarantee full connectivity of the network and to optimize the use of the global system capacity and the ratio between network performance and cost.
3) Definition of an "*external world*" interfacing policy that allows each UCOL station to satisfy the specific needs of its area both qualitatively and quantitatively by respectively supporting simultaneous connections of disparate types of traffic and providing a high number of physical attachments (up to 480 interfaces).

UCOL relies on a star topology which allows passive interconnection of stations which are not integral parts of the transmission medium thereby reducing demands on station reliability. One of the major requirements on UCOL is "*flexibility*" that is the capability to support different types of services and to facilitate the implementation of future and unknown services. It is important to point out here that the integration of any number of services and the flexibility in bandwidth allocation with agile mixing of high and low bandwidth users has led to recognize the need to operate a TDMA scheme on each optical frequency. Frequency domain switching between channels is therefore exploited essentially to obtain inexpensive adaptive network reconfiguration without hardware modifications.

The project therefore has set out to demonstrate that the bandwidth and the optical frequency tunability offered by coherent optics can be conveniently exploited to achieve a universal network able to connect pre-existing networks and to support wideband communication. As a consequence, multichannel networks can be considered as a natural evolution of already existing high speed LANs and not as new systems not open to the communication with pre-existing networks.

The latter point has led to consider within this project the results of the rapid progress made in the development of *Asynchronous Transfer Mode* (ATM) techniques. From the technical standpoint ATM emerges as a universal transfer technique with no special functions tailored to particular services, and is therefore a relevant reference point for a system such as UCOL which is aiming at providing a transparent communication world.

UCOL could carry out a primary role in the currently emerging scenario where private networks are interconnected, in a city-wide area, by means of MANs and in wider areas by the public B-ISDN, provided it allows an easy interworking with the above networks. These considerations have led to the adoption of ATM techniques with the specific aim to insure that the objectives of the project are met, i.e.:

- to define a network architecture independent of the specific services;
- easy interworking with public and local networks that will be based on ATM techniques, thereby minimizing the gateway functions;
- to avoid non-open solutions.

The most apparent impact of the introduction of the ATM techniques in UCOL is related to the way the information is carried through the network. The information flow is segmented into a number of fixed-length cells. Each cell is made up of a header field (5 bytes long) for control purposes and an information field (48 bytes) containing user data. The asynchronous nature of ATM is meant to be a non deterministic ordering of different information streams in an ATM multiplexed channel. There are no implications on the transmission systems used to carry ATM information. In UCOL the transport mechanism along the fiber is a framed TDM, therefore the ATM cells are inserted in time slots within the frame structure.

The first classification of services supported by the UCOL network, based on the ways the information is generated by different sources, can be represented by the following two types of traffic:

- *Continuous Bit-stream Oriented* (CBO) *services* (e.g. voice, video)
- *Bursty oriented services*, (e.g. data, variable rate video codec)

Owing to the different nature of the information sources, different interface functions must be performed to adapt the external traffic to a common internal transfer mode. CCITT I.321 recommendation defines a layered protocol architecture, where the adaptation layer is responsible to make the ATM layer independent of the type of service. Since the ATM layer *Protocol Data Unit* (PDU) has a fixed size, the adaptation layer must perform segmentation and reassemble functions and must guarantee the correct order of *Service Data Units* (SDUs) provided by the higher layer. It goes without saying that future services and terminals will not need adaptation functions because they will be conceived according to the user-network interface (not yet standardized).

The mapping of the Adaptation sub-layer and the ATM sub-layer in the UCOL node is presented in Fig. 3.1, where three types of services are offered:

1) *Connection Oriented* (CO) isochronous,
2) CO non-isochronous,
3) *Connection-Less* (CL) non-isochronous.

In the Adaptation layer the assembling/disassembling functions are outlined while the ATM sub-layer has been assigned for the basic function of handling the identifiers of connections, i.e. the *Virtual Channel Identifier* (VCI). Further

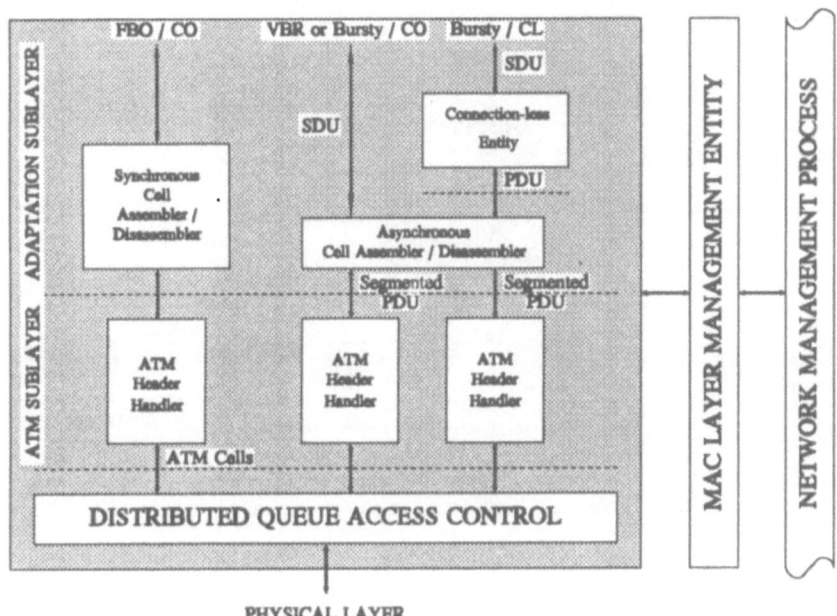

Fig. 3.1 - Structure of Adaptation and ATM layers

functionalities must be defined (service priority, error detection, synchronization recovery etc.) and split among the two layers.

The proposed architecture is mapped in the *Medium Access Control* (MAC) sublayer, the lower part of the Data Link layer. The CO isochronous services require establishing a connection before the actual information exchange starts. A VCI value must be assigned to the call and this value, as cell header, will unambiguously identify each information field of this connection carried through the UCOL network. Furthermore, the requested bandwidth must be guaranteed throughout the connection period. The signaling protocol is a version of the Q.931 recommendation (Fig. 3.2) enhanced with such functionalities as handling of a multichannel network, handling the call parameters according to the ATM characteristics, signaling for distributive services, etc.

The CO non-isochronous services basically refer to data communications, as LAN interconnection, workstation network, etc. A possible functional profile as stated by ISO according to the OSI reference model is shown in Fig. 3.3.

The connectionless non-isochronous services need some specific consideration. The maximum packet length handled by the *Logical Link Control* (LLC) IEEE 802.2 protocols is about 9000 bytes and since the information field of the ATM cell is far shorter, the problem to segment the original LLC Protocol Data Unit (LLC_PDU) and provide a robust mechanism to univocally reassemble the packet at the destination side arises. One possible solution is to emulate a connection for the duration of the packet transmission, assigning the same VCI to all the segments

USER PLANE
CONTROL PLANE
END TO END LAYERS
NL=Enhanced Q.931
EMPTY
LLC type 1 / 2
ATM
PHY

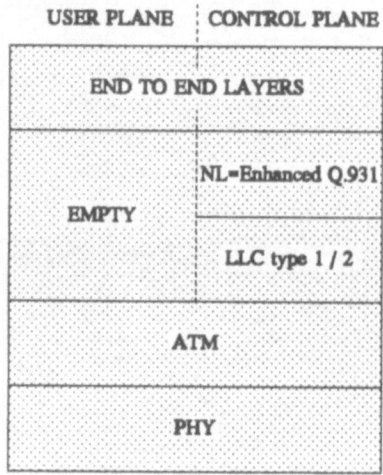

Fig. 3.2 - Profile of CO isochronous service

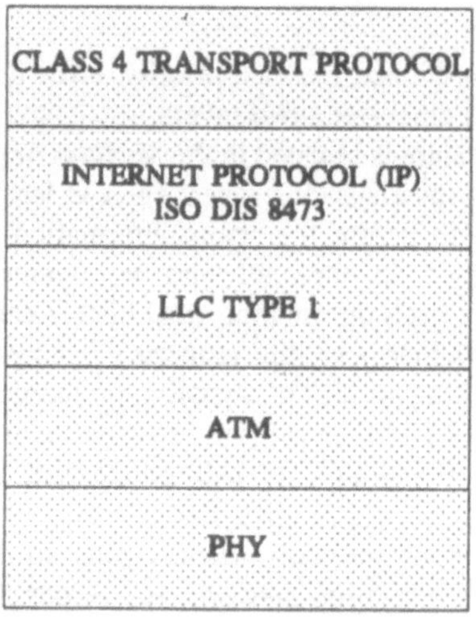

CLASS 4 TRANSPORT PROTOCOL
INTERNET PROTOCOL (IP)
ISO DIS 8473
LLC TYPE 1
ATM
PHY

Fig. 3.3 - Profile of CO data service

of a single packet. Fig. 3.4 shows the two phases of the CL service handling:

- building of the MAC_PDU (adding source and destination address)

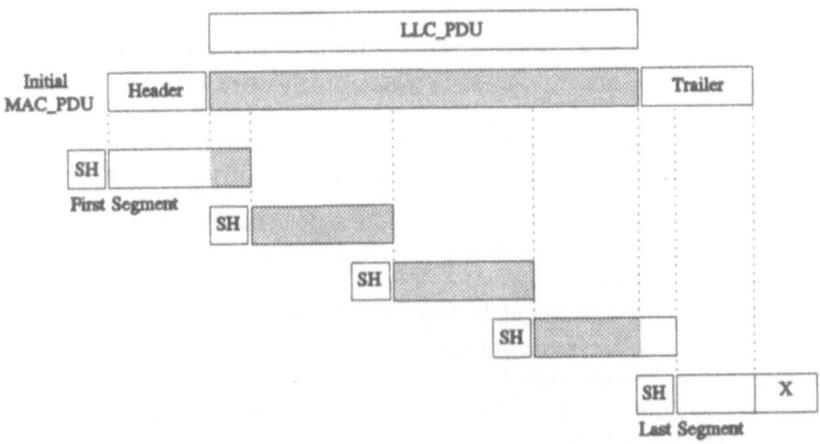

Fig. 3.4 - Handling of CL.service

- MAC_PDU segmentation (segment overhead provides the reassembling mechanism).

The lower functional block of the layered structure of Fig. 3.1 is represented by the *Distributed Queue Access Control* (DQAC). The access protocol must guarantee an ordered and collision-free access of data bursts to the star center, which is the only network resource shared among the nodes. It also must ensure an efficient use of the medium capacity available for information interchange among stations and a fair sharing of that capacity offering good *Quality Of Service* (QOS).

The fulfillment of the above mentioned requirements has led to the definition of the access protocol basic concepts that can be summarized as follows:

- the optimized signal multiplexing at the optical star
- the adoption of a frame based transmission scheme
- the *Queue Status* (QS) based access mechanism
- the prediction mechanism

It is important to stress that the access protocol functions are referred to a single channel and each entity handling a channel operates independently from all the others.

Optimized signal multiplexing means that data bursts coming from different network interfaces must be separated by a time gap negligible compared to the duration of the burst: the reason of this requirement is to be found in the need for an efficient exploitation of the available bandwidth. A necessary condition that is to be met in order to achieve optimized multiplexing is that every station has to know the exact time, referred to the optical star, at which its data bursts have to reach the star center. This, in turn, translates into the following conditions:

- a time reference signal has to be cyclically distributed to all the stations
- every station has to know in advance where to put its data bursts within the cycle
- every station has to anticipate the transmission of its data bursts by an amount of time equal to the propagation delay along the fiber.

The first condition is satisfied by the adoption of a frame based transmission scheme, since the cyclically transmitted frame synchronism can also be used as the star center reference time signal.

The second condition can be satisfied by providing all stations with the necessary information; the frame is basically composed of two parts, namely the Queue Status field and the Information field. The former has a fixed scheme, while the latter can be constructed on the basis of the information carried by the Queue Status in the preceding frame.

The third condition is satisfied during the Station initialization, when the measurement of the station distance from the star center is performed.

The adopted frame (Fig. 3.5) is 1 ms long and is made up of three parts:

- Frame Header: 648 bits
- Queue Status field: 28,416 bits
- Information field: 270,936 bits

The frame header is generated by one of the stations: if, for any reason, this station should go down, an appropriate procedure is defined so that a new station starts generating the frame synchronism.

The QS field is divided into 64 control slots, one for each station, according to a fixed transmission order: the first slot is reserved to station 1, the second to station 2 and so forth (this implies that, for the QS field, the second condition of optimized multiplexing is satisfied). The QS provides all stations with the information they need to decide which data slots they will use: it is important to note that the information carried by the QS field of frame T is used to arbitrate the access to the Information field of frame T+1: so, when a new frame starts, every station already knows which slots of the Information field it will use for that frame (thus satisfying the second condition also for the Information field).

The Information field is divided into 426 data slots each able to contain one ATM cell; the channel net bit rate is $426 \times 48 \times 8 \times 1000 = 163.584$ Mb/s. This capacity is adequate to handle, if required, either the service payload carried by the second rank of the *Synchronous Digital Hierarchy* (SDH) or the likely rates currently considered for video services. Furthermore this value of net capacity exceeds the maximum rate of any underlined individual broadband service currently expected to appear in the medium term.

Each control slot in the QS field carries information which is basically used to arbitrate bandwidth reservation (RSBW and RQBW fields) and data slots allocation (P0QS-P4QS and P1PR-P4PR fields). The arbitration is not carried out by a

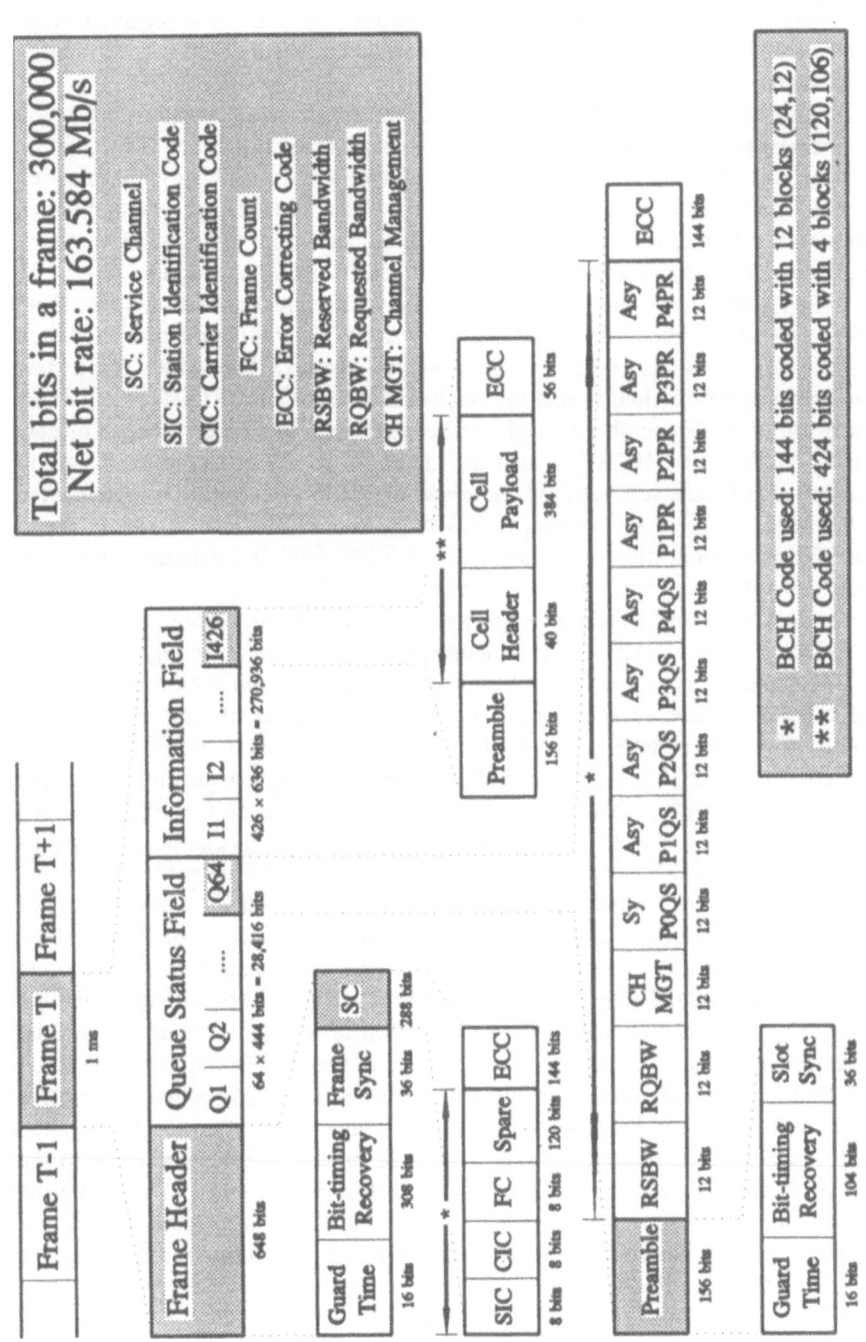

Fig. 3.5 - UCOL frame structure

centralized entity, but is distributed among the stations; the QS is received by all stations and elaborated concurrently in each of them. The univocity of arbitration is guaranteed by suitable algorithms.

An important aspect of UCOL is the use of the same access mechanism for both synchronous and asynchronous traffic: the only significant difference is that, before a synchronous connection is accepted, a sufficient amount of bandwidth must be granted. That is, a mechanism must be provided that reserves a given number of slots on each frame, for all the duration of the connection: this assures that new synchronous calls do not subtract bandwidth to calls already in progress. In UCOL, bandwidth reservation can be performed only for one connection at a time; synchronous traffic is handled as traffic with priority P0, while priorities P1-P4 are used for asynchronous traffic. When a station requires bandwidth for a new (synchronous) connection, it fills the *ReQuested BandWidth* (RQBW) field of its control slot with the number of cells it intends to use (in case of absence of new connections requests, RQBW is filled with zeros). If its request is satisfied, starting from the next frame, its *ReServed BandWidth* (RSBW) field will be incremented by the number previously written in RQBW, while RQBW will be again set to zero. When a connection is dropped, the RSBW field is decremented by the number of cells (per frame) that the connection was using.

Data slots allocation procedure is quite simple: after all the stations have received the QS they begin to elaborate it concurrently. Starting from the highest priority (P0), slots are reserved one for each station (if, of course, the station has a cell to transmit) until either every station has obtained a number of slots equal to the number of cells to be transmitted (that is the number written in the P0QS field) or the frame has been filled up. Only after all slots of a given priority have been assigned, slots of next priority will undergo the allocation procedure. Since requests with priority P(i) are served before those with priority P(i+1), the handling of traffic of a given priority is not influenced by the presence of lower priority traffic.

The prediction mechanism has been devised for a more efficient exploitation of the available bandwidth: the P1PR-P4PR fields are used to reserve additional slots, besides those indicated in the P1QS-P4QS fields, according to an appropriate prediction algorithm. These slots will possibly receive all or part of the cells that arrive in the queues in between the transmission of the QS on frame T and the transmission of data on frame T+1. Different possible solutions have been evaluated for the prediction mechanism: the selected method is called "*exhaustive C*". After all Data Slots have been allocated according to the requests written in the P0QS-P4QS fields, additional slots are reserved to each station, according to the requests in the prediction fields (P1PR-P4PR): this second form of allocation is repeated more times, until all slots in the frame have been exploited. These additional slots are not tied to a particular priority: in fact, during the transmission of the next frame, slots will be used to accommodate the highest priority cells available at the moment.

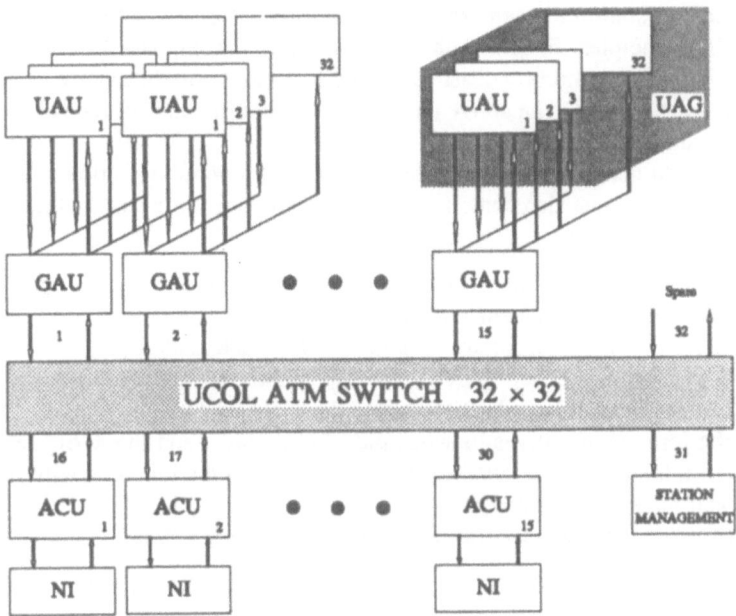

Fig. 3.6 - Architecture of an UCOL Station

3.1 UCOL Station Architecture

The UCOL station architecture is shown in Fig. 3.6, where three different sections can be identified.

The upper part represents the UCOL-external world interface. Each user terminal equipment accesses the network through a dedicated interface called *User Access Unit* (UAU). The lower part represents the station interface to the optical star. There is one *Access Control Unit* (ACU), performing the already mentioned DQAC protocol, for each optical *Network Interface* (NI). The middle section is the internal switching system, which provides the interconnection of the overall component units of the station.

The key issues examined during the preliminary definition of the station architecture are:

1) Switching capability in terms of blocking probability, introduced delays, priority handling capability, number of simultaneous connections, broadcast and multicast support;
2) modular extension;
3) easy adaptability to the emerging public and private communication scenarios.

The proposed architecture is built up around the basic concept that the switching functions are performed by means of ATM cells routing. This fact implies that information (signalling, user data, management) within UCOL is exchanged, between one station element and any other element of the network, according to the ATM principles: fixed length cells, switching accomplished on virtual circuit basis. A multicast and broadcast capability is offered both internally to the station and at network level. The above characteristics require the deployment of an ATM switch within each station.

In order to achieve such a flexible and powerful communication scheme, each station element must also contain a common communication stack which is formed by the following layers:

- at the highest layer the management protocol, to assure message exchange between peer *Management Entities* (ME);
- LLC, for the provision of services of the OSI Data Link layer;
- *ATM Adaptation Layer* (AAL), to implement the adaptation functions and procedures needed in order to translate the requests coming from the LLC layer (in the form of primitives between the LLC sublayer and the MAC sublayer, as stated in the IEEE 802 recommendations) into information units with the size of the cell payload (48 bytes);
- ATM layer, that includes the cell header handling functions.

Each of these layers has its respective *Layer Management Entity* (L_ME). Each L_ME will be designed in order to perform the tasks specific for its layer, but it must also have a part that is common with the other L_ME, whose function is to exchange management information with its hierarchically superior management entity, the *Station Management Entity* (S_ME), in order to provide monitoring, control and coordination of all the managed objects of each layer.

The station switching system is built up by an ATM switch and several parallel *Peripheral Buses* (PBs). The former is a multistage ATM switch, where each stage is an 8×8 switching element running at 150 Mb/s per link. For a fully equipped station (including 15 NIs) a 32×32 switch is required. The PBs, where multiple UAUs can be connected, act as concentrators of all user equipments requiring bandwidth far less than the maximum capacity of one single 150 Mb/s link. Each PB can arrange a maximum of 32 UAUs (a cluster of UAUs connected to the same PB forms an *User Access Group*, UAG) and up to 32 PBs can be inserted in a single station.

The function of parallel to serial and serial to parallel conversion is carried out by the *Group Access Unit* (GAU).

In order to guarantee that UCOL be open towards the external world, UCOL management will be OSI compliant. This means that, from an external point of view, UCOL will be able to exchange management information with all systems adopting the OSI philosophy, in terms of management. The requirement of an OSI compliant management affects only the external behavior of UCOL; the internal management architecture does not depend on it. Therefore, although UCOL management is functionally a single entity, it is in fact distributed over a hierarchical structure reflecting the internal UCOL architecture.

It is possible to identify three hierarchical levels in the UCOL management structure, as it is shown in Fig. 3.7.

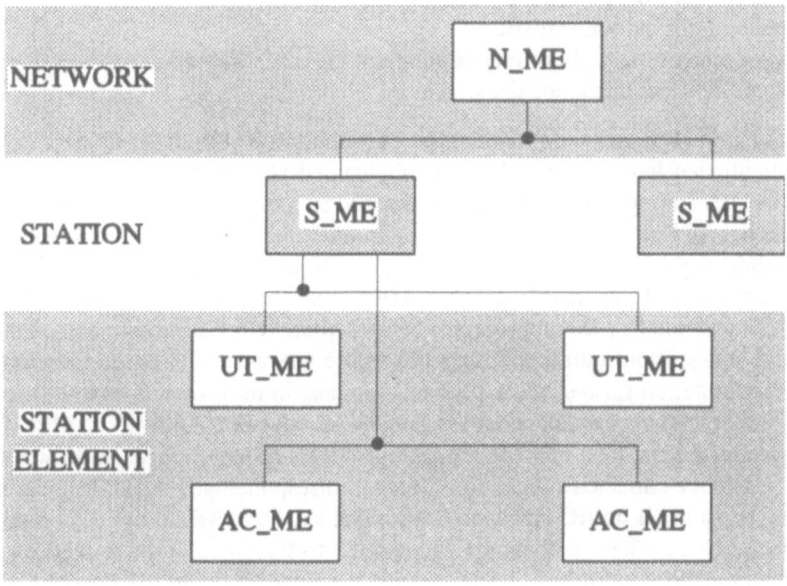

Fig. 3.7 - UCOL management structure

The highest level (network level) consists of the *Network Management Entity* (N_ME). This functionality will reside in an external device (a workstation or a personal computer) and will be responsible of the supervision and control of the entire UCOL network. This external device will be physically connected through a dedicated LAN (as shown in Fig. 3.8) or directly via an UAU to one of the UCOL stations. As all UCOL stations have the same potentialities, the identity of the connected station is functionally not relevant. The functions of N_ME will be those of general interest for UCOL:

- management of the ordinary and exceptional behavior of the network;
- gathering of network statistics;
- management of network reconfiguration in case of traffic congestion, faulty conditions, on line extensions.

The second hierarchical level (station level) is composed of the *Station Management Entities* (S_MEs), replicated in the management boards of each station. S_MEs are responsible, towards the N_ME, of the management of all the resources of the station, of the optimization of their behavior, of their representation in the *Station Management Information Base* (S_MIB), of the consistency of the values of their parameters, etc., but also of the traffic "routing" over the optical network (the choice of the channel on which a connection has to be established or asynchronous traffic has to be placed). As for their common parent N_ME, also the S_Mes will feature the functions:

- management of the ordinary and exceptional behavior of the station;
- gathering of station statistics;
- management of station reconfiguration in case of traffic congestion, faulty conditions, on line extensions.

It can be seen that the functionalities are the same at both levels and, as it will be mentioned later, this similarity is maintained also at the next level. The basic difference among them lies in the functional context that they are called to supervise and control. In order to better clarify this concept, a short example on statistics acquisition is given. Whenever the N_ME is requested (by the network operator) to build up statistical reports (e.g. histograms) on the channels load, the N_ME will request the appropriate S_MEs (that is the least number of stations whose optical configuration covers the whole spectrum of frequencies) to gather load information for a specified list of channels; in turn, each S_ME will forward the request to the appropriate *Access Control Management Entity* (AC_MEs). These will activate local load-collection procedures that, on a periodical basis, will return the respective collected acquisitions to their corresponding S_ME. The information received from all its AC_MEs, are elaborated by the S_ME to build up reports on a station basis. These will finally be reported to the N_ME who will take care of properly elaborating it in order to provide the final results to the operator. During the visualization session, the *Man-to-Machine Interface* (MMI) will support the N_ME application functions to provide on-line facilities that help and guide the operator in a correct and exhaustive analysis of the results.

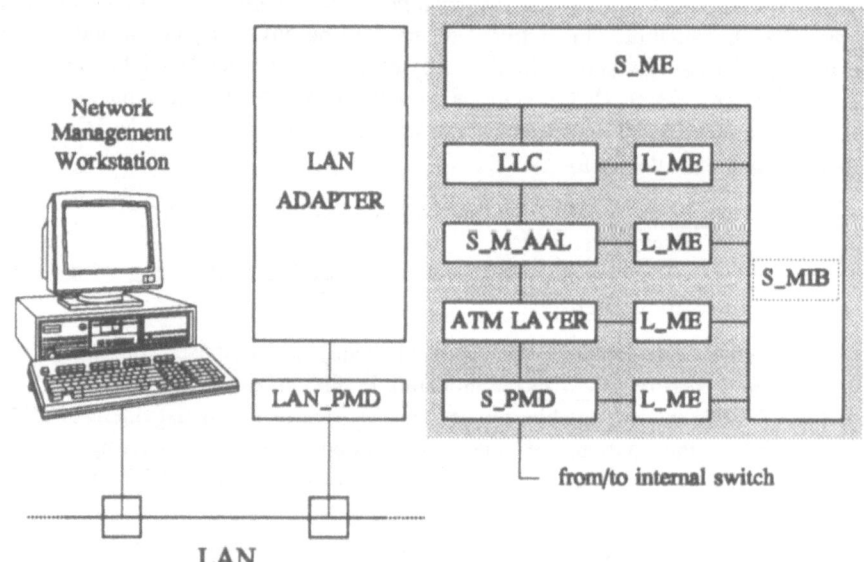

Fig. 3.8 - Structure of the Station Management Unit

At the lowest level (station element level), there are two distinct types of management entities which will be generically addressed as *Station Element Management Entities* (SE_MEs). These are respectively the *UCOL Termination Management Entity* (UT_ME) and the above defined AC_ME, which are placed in different boards of the stations. Like the higher levels entities, the SE_MEs are responsible for the management of ordinary and exceptional behavior of their own environment, for gathering of local statistical data and for the management of local reconfiguration in case of resource congestion, faulty conditions, on line extensions or higher level explicit requests.

Figure 3.8 shows the structure of the *Station Management Unit* (SMU) and its relationship with the N_ME, in the case where this is resident on a workstation and connected to UCOL station by means of a LAN. In this situation, it is necessary to have a LAN adapter. The SMU is composed of the S_ME, together with the S_MIB, and by the communication layers that allow the management board to dialogue with the other boards of the station; these are LLC, *Signalling & Management ATM Adaptation Layer* (S_M_AAL), ATM Layer and *Station Physical Medium Dependent* (S_PMD). Each of these communication layers has its respective L_ME. ATM cells coming out of the S_PMD go to the other boards through the ATM switch of the station.

The structure of the other units of the station is substantially similar to that of the SMU: in fact, they have the same communication stack towards the internal ATM switch, as shown in Fig. 3.9 (Access Control Unit) and in Fig. 3.10 (*UCOL Termination*, UT).

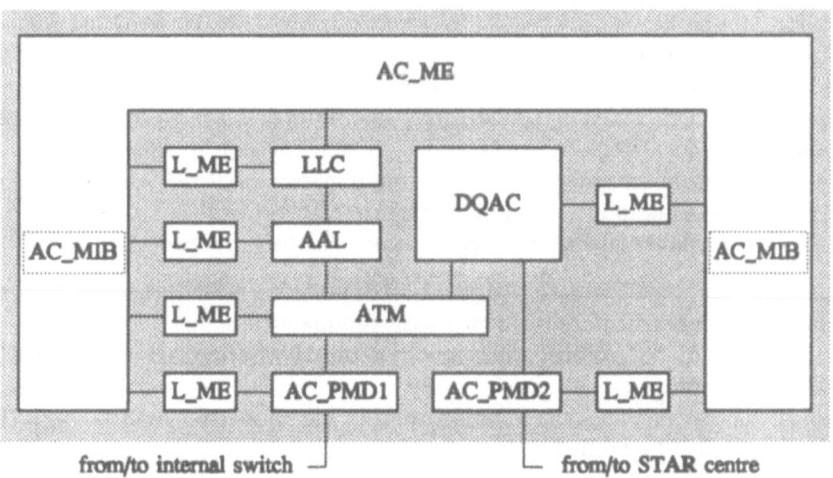

Fig. 3.9 - Structure of an Access Control Unit

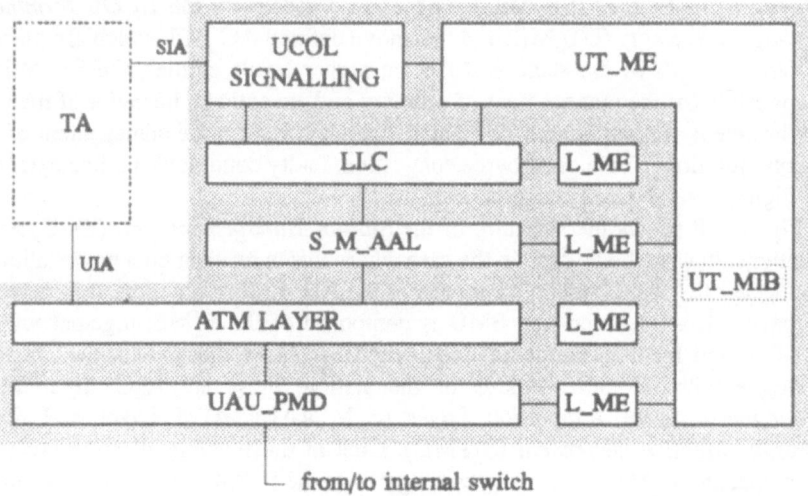

from/to internal switch

Fig. 3.10 - Structure of the UCOL Termination

The internal structure of the ACU (Fig. 3.9) is completed by other elements that identify the processing of the ATM cells through the ACU, as far as communication between stations is concerned. The cells coming from the AC_PMD1 layer (where "1" identifies the Physical Medium Dependent towards the ATM switch), are filtered by the ATM layer and delivered to the DQAC layer and finally through the AC_PMD2 layer (where "2" identifies the Physical Medium Dependent towards the optical interface) sent to the optical star; in the reception phase, the flow is exactly the reverse one.

Each UAU is functionally divided into two blocks, as shown in Fig. 3.11: the *Terminal Adapter* (TA) and the UT (Fig. 3.10). The aim of this splitting is to maintain a logical separation of the functions integrated in the UCOL environment from those relating to interfacing with the external world.

The TA, as shown in Fig. 3.12, carries out three basic functions:

- termination of the *Terminal Equipment* (TE) protocols with the separation of the control information from the actual user information;
- adaptation of the control information to the form required by the UCOL Signalling entity;
- adaptation of the actual user information to the format required by the ATM layer provided by the UT.

Signalling information will pass from the TA to the UT across the *Signalling Information Access* (SIA) interface, while data flow from the TA to the UT across the *User Information Access* (UIA) interface.

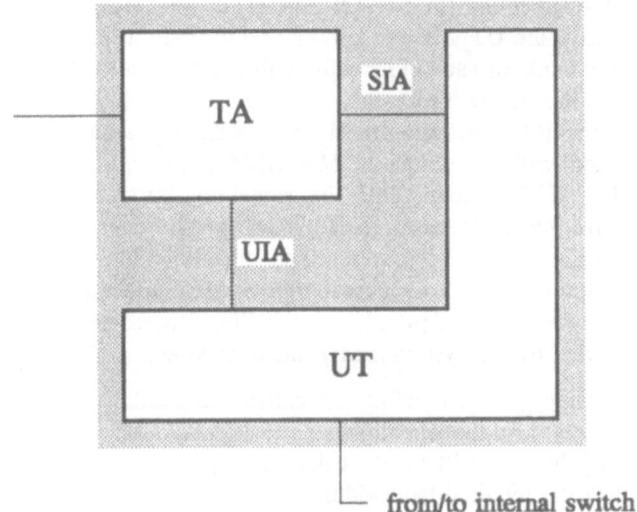

from/to internal switch

Fig. 3.11 - Structure of an User Access Unit

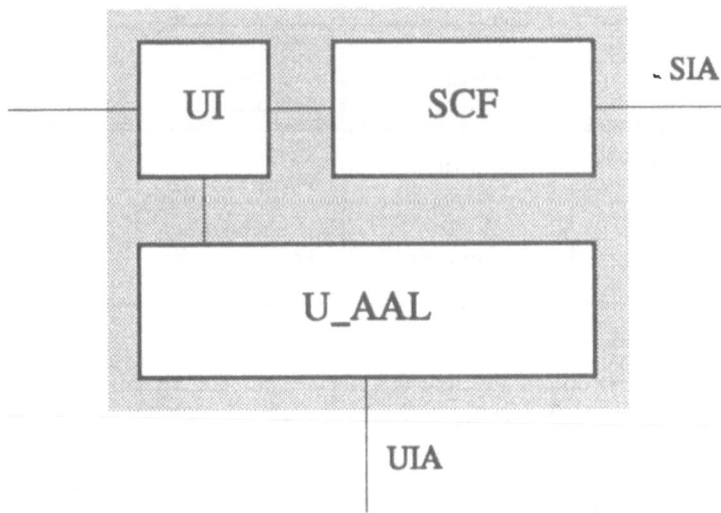

UIA

Fig. 3.12 - Structure of the Terminal Adapter

Besides the usual communication stack towards the internal ATM switch, there is another functional unit in the UT: UCOL Signalling (see Fig. 3.10). It receives from the TA the signalling (in a UCOL format) coming from the external world, and uses it to give the UT_ME the information to perform its functions.

The implementation of each block will be different for all UAUs, depending on the kind of traffic to be processed. The aim of the *User Interface* (UI) is substantially to split the incoming traffic in data and signalling information. Data information, then, will go through the *User ATM Adaptation Layer* (U_AAL) and will reach the ATM switch, while the signalling information will cross the *Signalling Convergence Functions* (SCF) where it will be made "*understandable*" to UCOL Signalling. Once it has received signalling information, the UCOL Signalling block will have to interact with the S_ME. In fact, in order to establish a connection between two end points, the signalling process needs information and operations handled by the system management, such as:

- common channels between calling and called end points;
- channel to be used by the signalling;
- assignment of the VCI/VPI for the cells carrying signalling information;
- assignment of the required bandwidth;
- assignment of the VCI/VPI for the actual user data transfer.

The signalling protocol will be based on the CCITT Q.931 enhanced with the needed functionalities capable of supporting distributive services, multimedia communication and multichannel operations.

Table 3.1 - Characteristics of CO and CL services.

Service	Basic Communication Unit	Addressing Information	Routing Identifier		
Connection Oriented (CO)	Call	Out of band Private or public	VCI		
Connectionless (CL)	Packet (Message)	In band Private or public	VCI	MID	Address

The UCOL network architecture described is based on ATM which, as such, is a connection oriented technique where virtual circuits are established on a connection basis. Table 3.1 shows how the call represents the basic communication unit for CO services.

Each call is delimited by a set-up and a tear-down phase (call control process). Between these two phases, the actual communication takes place always referring to the same virtual circuit identifier. An important characteristic of CO services is the negligible set-up time with respect to the overall duration of the actual exchange of user information. The establishment of a call is typically realized through three consecutive steps:

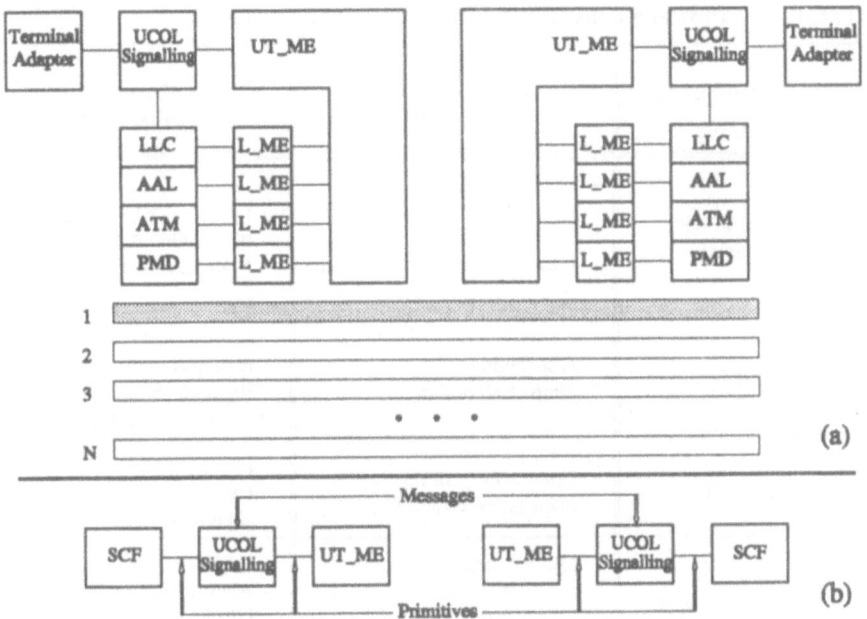

Fig. 3.13 - a) Functional entities involved during the signalling process; b) Primitives and Messages

1) signalling resources identification (channel, VCI)
2) bandwidth reservation (performed by the access protocol as described in the previous paragraph)
3) end-to-end signalling. Fig. 3.13 shows the functional entities involved in this process and the types of mechanisms used to perform communication among them.

Steps 1 to 2 can be reiterated a number of times in case bandwidth reservation fails and more channels (between calling and called user interfaces) are available. The selection of the channel represents the main routing function in UCOL (in fact the transmission mechanism, on each channel, is inherently broadcast). A number of different allocation schemes have been analysed (their pros and cons are summarized in Table 3.2):

a) *Optimum Load Balancing* where the request is routed over the less loaded channel;
b) *First Fit* where the request is routed over the first channel with sufficient band;
c) *Best Fit* where the request is routed over the channel that presents an amount of free bandwidth closest to the requested one
d) *Connection Based* where preferential channels, common to calling and called stations, are selected.

It is worth highlighting the basic concept of reallocation which is intended for exploiting the multichannel feature of the optical network. Reallocation is used to

Table 3.2 - Pros and cons of the analysed allocation schemes.

Channel Selection Criteria	Advantages	Disadvantages
Optimum load balancing	- High probability to satisfy small to medium bandwidth services requests - Little probability to encounter contention situations	- Increase complexity of the reallocation mechanism - More complex algorithm
First Fit	- Fast routing - Simpler algorithm	- Increase complexity of the reallocation mechanism - Random connections distribution on the channels
Best Fit	- Optimum use of bandwidth (increase of efficiency)	- Increase probability to encounter contention situations
Connection-Based	- Network setting as a response to traffic demands - Decrease complexity of the reallocation mechanism	- Algorithm requires more information

optimize the distribution of traffic over the channels and consists in "moving" one or more connections from one channel to another.

Assignment of the VCI is the second important function to be carried out when a new connection has to be established. Since in ATM based networks VCI is the only parameter that distinguishes one connection from the others, in each call the uniqueness of the VCI assignment must be guaranteed. Fig. 3.14 shows the path of the information cells during the connection between UAU X of station A and UAU Y of station B. The points where VCI based switching functions are accomplished are the ATM switches in the stations. It is possible to observe that two distinct domains exist within which the VCI uniqueness must be guaranteed: the Network domain and the Station domain. This principle is supported in UCOL by the capability that the ATM switch has to translate VCIs of connection segments of one domain to the corresponding segments in the other domain.

For connectionless (CL) services, the basic communication unit is a single packet with variable length, whose size can extend up to around 9 kbytes. It is apparent that the mechanism adopted for CO services, that is, setting up a virtual circuit for each communication unit, is not applicable for CL services. In this case the call control procedures would last much longer than the duration of the actual user information exchange, producing an unacceptable service degradation, particularly when fast data transfer is requested. UCOL provides a flexible solution capable of coping with different user requirements. The basic application scenarios for CL data services are:

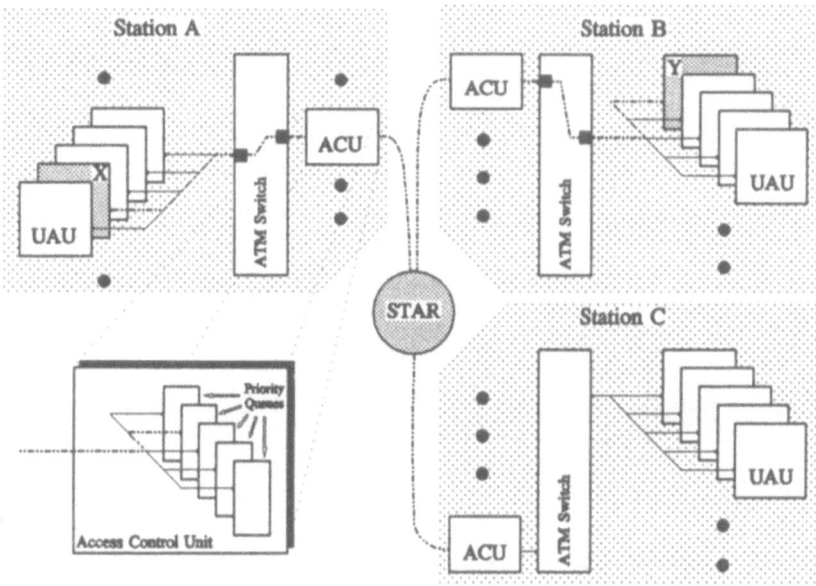

Fig. 3.14 - Information cells path during a connection between UAUs X and Y

- LANs interconnection
- Distributed Processing Systems.

The application scenarios for CL data services are shown in Fig. 3.15, illustrating UCOL capability of interconnecting LANs and TEs.

UCOL provides network access to the users by means of dedicated interfaces (UAUs); these can be built with gateway or bridge functions, allowing interconnection of heterogeneous and homogeneous LANs respectively. In the first case the interconnection is realized at level 3 and UAUs are different from each other since each of them has to perform conversion between the specific LAN protocol (e.g. DECNET, XNS, TCP/IP) and UCOL own protocol. Referring to Fig. 3.15a, if the DECNET LAN wants to communicate with the XNS LAN, the first UAU has to provide DECNET/UCOL protocol conversion, while the second UAU performs the conversion UCOL/XNS.

The interconnection of homogeneous LANs can be made at level 2 (Fig. 3.15b), with the UAUs performing bridge functions; this allows the capacity of LANs to be extended, both in terms of number of users and of geographical extension. In this case the UAUs are similar to each other and they do not have to provide protocol conversion.

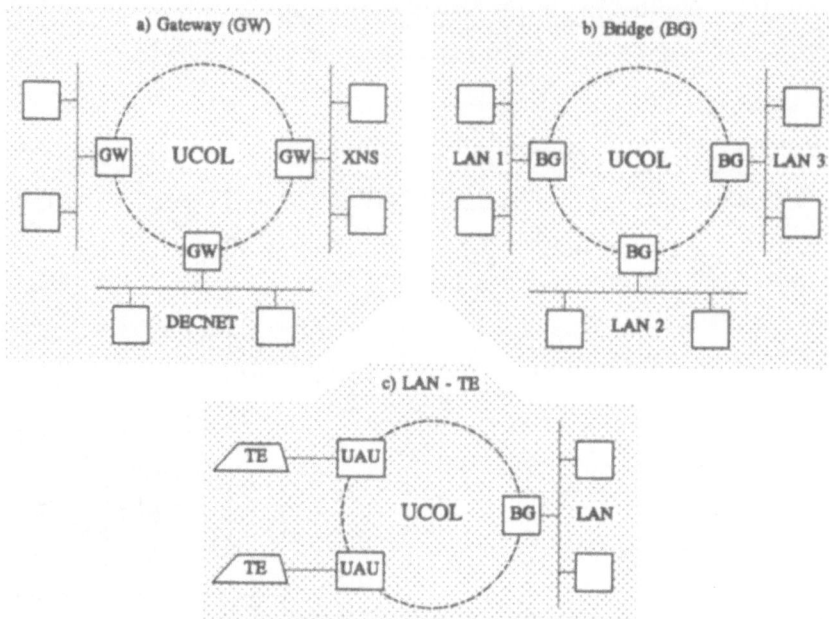

Fig. 3.15 - Application scenarios for connectionless data services

Figure 3.15c refers to the case in which a workstation or a terminal equipment is directly connected to UCOL.

Figures 3.16 and 3.17 show the functional diagram of two UAUs with gateway and bridge functions, respectively. The shaded part is the same in both cases and represents the common UCOL CL stack which allows any kind of connectionless service to be transported through the UCOL network, while the remaining part is service dependent.

The common UCOL CL stack provides services to the upper layer as a response to a primitive which, in case of gateway is a standard primitive between sublayers LLC and MAC (MAC_DATA_REQUEST with parameters MAC_SDU, SA, DA, QOS).

In the case of a bridge the routing (operating at level 2) analyzes the *Source Address* (SA) and the *Destination Address* (DA) fields, builds up the new SA and DA and then communicates with the ATM stack by means of a primitive similar to the first one. The bridge approach is independent of any higher layer and allows homogeneous systems to be transparently connected. Also, it allows a greater packet throughput than a gateway approach.

The basic characteristic of CL services is that they rely on broadcast transmission thus relieving the sending station from setting up path(s) towards the destination(s).

The solution adopted in UCOL is based on the same principle and is realized by introducing the *Service Virtual Network* (SVN). Those applications connected through dedicated UAUs to a number of stations that need to exchange

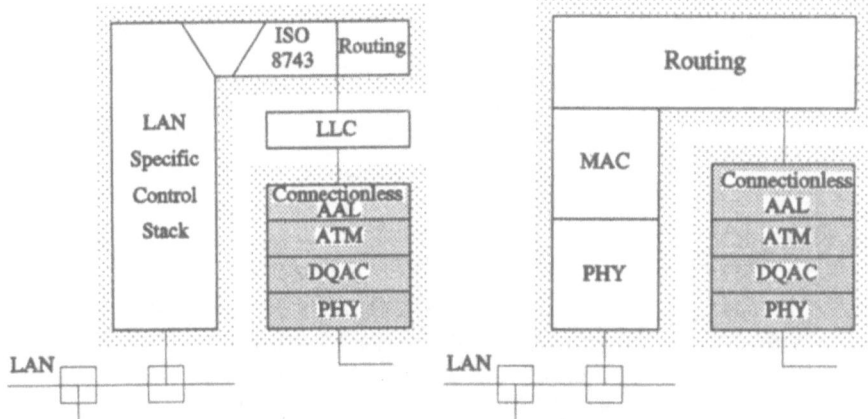

Fig. 3.16 - UAU with gateway functions   **Fig. 3.17** - UAU with bridge functions

information, constitute a SVN. The basic requirement for a SVN is that all its station members must be tuned over the same optical channels. The mechanism that will be used in order to physically select one SVN instead of another is based on setting up semipermanent associations between UAUs and optical channel. The SVN is identified by a number which must be unique within the network and all UAUs which are members of that SVN will always use that same number. This SVN identifier will be inserted in the VCI field of the ATM header and will be used by all members of that SVN exactly in the same way:

- at the transmitting side, the UAU formats the ATM cells and inserts the SVN identity number into the VCI;
- the UAU injects the cells on the Peripheral Bus towards the switch;
- upon reception of these cells, Switch Connection Control translates from VCI to network-switch port and relays them unchanged in that direction (the cells reach the ACU);
- at the receiving side and upon reception of these cells, Switch Connection Control translates from VCI to user-switch port(s) and relays them unchanged in that direction (the cells reach the destination Peripheral Bus).

Since within the same SVN a unique VCI is used, the CL communication units (packets) are identified by means of the *Message Identifier* (MID) which is a field added within the information payload by the segmentation and reassembling entity (see Fig. 3.18).

Finally, since both CO and CL services use VCIs, simple rules avoid VCI conflicts (e.g.: a field in the ATM header is used as service discriminator).

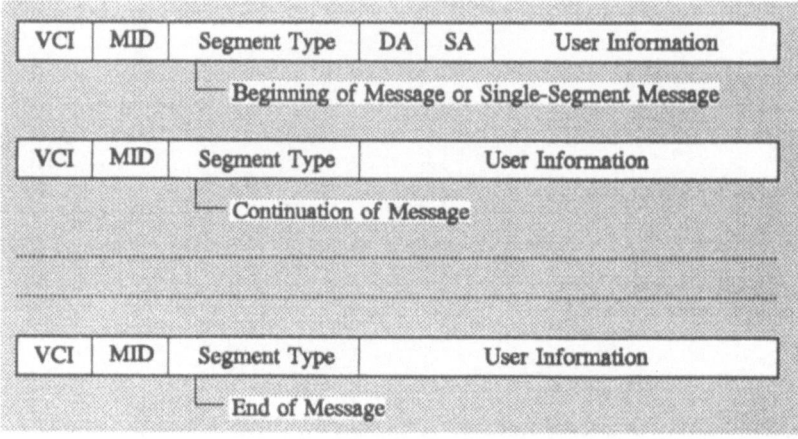

Fig. 3.18 - Structure of the CL communication units.

4 UCOL Optical Architecture

4.1 Introduction

S. Forcesi and E. Neri

UCOL is a network interconnecting a large number of *Network Interfaces* (NI), i.e. a Transmitter/Receiver pair; the link among Nis is performed by means of single mode fibers, , which are all connected to a central passive star, that acts as frequency multiplexer. All interfaces are able to communicate over a set of 20 FDM optical channels at the central nominal wavelength of 1535 nm. All NIs are frequency locked to a common reference source, each one realizing a 160 Mb/s information channel upon which information is carried by means of ATM cells that are time division multiplexed on the channel.

The available system gain allows a main configuration offering a connection, over a star radius of up to 10 km, for more than 1000 Nis. Each NI is directly connected to the central star in order to avoid additional losses due to local splitters/combiners.

The *Bit Error Rate* (BER) of the physical channel has been fixed at 10^{-6}: this increases the receiver sensitivity and makes it less vulnerable to adjacent channel interference. The channel quality is improved with *Forward Error-correction Coding* (FEC) techniques, which restore the BER to a value better than 10^{-13}.

The modulation scheme is the *Differential Phase Shift Keying* (DPSK) with a transmission gross rate of 300 Mb/s. The receiver is based on a polarization diversity scheme as it avoids the need of fast adjustment to the channel *state of polarization* (SOP).

The UCOL optical hardware is constituted (Fig. 4.1) of a number of Stations (up to 64), each one including up to 15 Nis, a central passive star and an auxiliary central coupler used for the distribution of the frequency reference. A *Reference Generation Block* (RGB), that provides all transmitters and receivers a frequency reference constituted by a set of equally spaced lines, is located in one of the stations: this reference comb is distributed by means of both the additional central coupler and the power splitters located in each station.

The RGB is composed of a *Mode-Locked Laser* (MLL), driven by a microwave signal at 3.6 GHz (corresponding to the comb lines spacing), followed by a polarization scrambler to enable SOP insensitive detection of the reference lines.

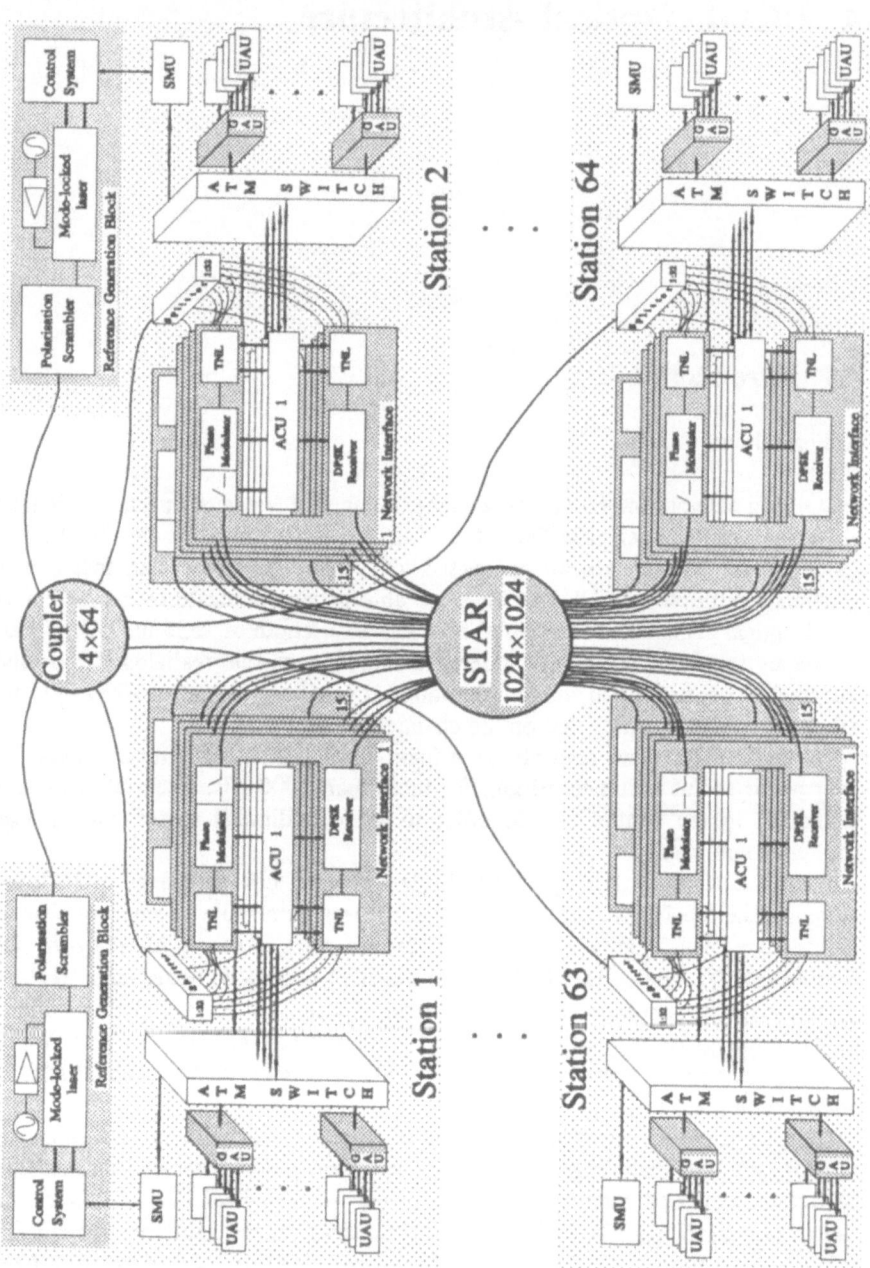

Fig. 4.1 - UCOL optical hardware

The fundamental optical component within each NI is the *Tunable Narrow-line Laser* (TNL), which is used both as carrier generator in the transmitter and as *Local Oscillator* (LO) in the receiver. The TNL can be locked, with a proper offset frequency, to any of the reference lines provided from the RGB by means of an *Automatic Frequency Control* (AFC) scheme. In the transmitter the TNL is followed by a phase modulator and an isolating switch; the latter is used to disconnect the transmitter from the network during its idle state. The receiver is composed of a balanced polarization diversity optical front-end (fed by both the LO signal and the incoming data signal), a low noise preamplifier and a delay-line DPSK demodulator.

As shown in the block diagram two dedicated stars are used for distributing signals and reference frequencies. This approach does not increase the total amount of fiberoptic cables required in the system and makes the selection and locking tasks to be performed in each station easier, as no interference with the data signal takes place.

4.2 Optical Receiver

S. Forcesi and E. Neri

The location of several optical carriers within the same spectral region, the bursty signal due to the *Time Division Multiple Access* (TDMA) and the imperfections of laser sources have imposed a series of constraints upon receiver design. The UCOL receiver, based on a polarization diversity optical front-end in a balanced configuration followed by a double arm DPSK demodulator, is shown in Fig. 4.2.

The received signal is combined with the LO signal, coming from the TNL, in a 3 dB directional coupler; this mixed signal is then split into orthogonal polarization components by means of two TE/TM splitters and detected by four photodiodes. The two pairs of balanced outputs corresponding to the same polarization are subtracted to suppress intensity noise; the two output signals feed the two arms of the demodulator.

Each arm is composed of a low-noise preamplifier based on the current-current feedback scheme, an *Intermediate Frequency* (IF) amplifier and filter, an *Automatic Gain Control* (AGC) module and a delay line demodulator. The two demodulated signals, each one corresponding to its polarization component, are added and then sent to a threshold comparator. The clock recovery function is accomplished by a dedicated circuit designed for fast acquisition.

A theoretical analysis of the receiver BER has been performed in the context of an ideal single channel system: the receiver sensitivity is about -61.5 dBm in order to obtain a BER of 10^{-6} with a bit rate of 300 Mb/s. This figure is reduced by a number of impairments which have been carefully evaluated, like:

- phase noise of both transmitting and LO lasers;
- *Relative Intensity Noise* (RIN) of the LO laser;
- Intensity modulation due to the active phase modulation process;

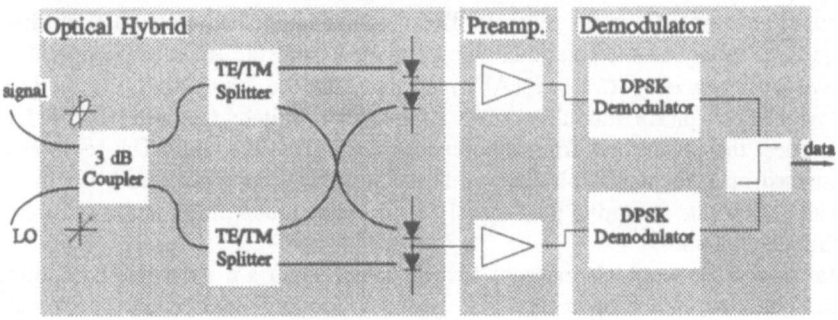

Fig. 4.2 - Block diagram of the UCOL receiver

- *Amplified Spontaneous Emission* (ASE) of the active phase modulator;
- shift of the IF due to the AFC
- imperfections in the physical realization of the optical components (insertion loss, unbalance and TE/TM extinction ratio);
- *Adjacent Channel Interference* (ACI) and *Co-Channel Interference* (CCI);
- photodiodes responsivity;
- preamplifier thermal noise;
- baseband and IF filtering;
- jitter of the recovered clock.

This degrades the receiver sensitivity to -50.9 dBm, that is a value which still allows for a link margin of 3.0 dB.

It should be emphasized that a BER of 10^{-6}, a relatively high value for MAN applications, has been selected in order to increase the system gain and to avoid the effect of ACI.

4.3 Forward Error-correcting Coding

S. Forcesi and E. Neri

Due to the relatively high BER and in order to allow the implementation of the unique frame adopted, error correction techniques have become necessary.

The application of error coding in UCOL has taken into account both frame format and the adoption of ATM, which set a number of constraints on the coding.

The information on each individual optical channel is organized in periodic frames of constant length (1 ms). As explained in Chapter 3, the frame format (Fig. 3.5) uses three major fields:

- frame alignment field;
- Queue status field, that basically carries the information related to packet priority and number of packets waiting for transmission in each individual node queue;
- information field, where each slot can accommodate an ATM cell of the format 48 + 5 bytes.

It is important to realize that the performance of the system is not determined by the basic BER but by the loss rate of these three fields. For instance, an incorrect QS field may cause the loss of the entire frame and therefore system requirements call for a probability of receiving a wrong QS field much less then 10^{-6}; for UCOL a value of 10^{-14} has been adopted.

Considerations on the management of synchronous traffic in addition to the bursty transmission mode of the network and the bit rate lead to a selection of a FEC technique based on block codes. Using a (24,12) Extended Golay Code for the QS field and a (120,106) BCH code for both the ATM cell header and the ATM data field, the following results have been found:

- QS field error probability: 1.3×10^{-19}
- ATM cell loss rate: 0.8 cell/Yr
- ATM cell with errors in data field: 15 cells/Yr

It should be stressed that the BER corresponding to an ATM cell loss rate of 15 cells/Yr is about 10^{-14} allowing for a future increase of demanded quality.

4.4 Channels Allocation Plan

S. Forcesi and E. Neri

Although the optical bandwidth available in the 1500 nm window is greater than 10 THz, allowing in principle the allocation of a large number of channels without any interference, limitations due to basic optical blocks makes the allocation plan a tricky issue. The main constraints derive from the limited tunability range of the optical sources and from the frequency reference system. The current narrow linewidth lasers have a practical tuning range of about 1 nm, but uncertainties in their central wavelength cut out the useful range. The reference system is based on the generation of a frequency comb, characterized by the number of generated lines and by their spacing; this reference comb is distributed to all the optical sources. The reference frequency comb is generated by mode-locking of an external cavity semiconductor laser. This device is currently able to produce a limited number of lines spaced by some GHz. Therefore the available bandwidth is rather small and, in order to allocate as many channels as possible, the optical channels are to be very close to each other.

The UCOL allocation plan is shown in Fig. 4.3; in the upper side the complete channeling scheme is detailed, with the reference lines (dotted lines) labeled with capital letters and the optical channels (solid lines) labeled with numbers. In the lower side a detail is enlarged, in order to show the LO frequencies and the channels spacing.

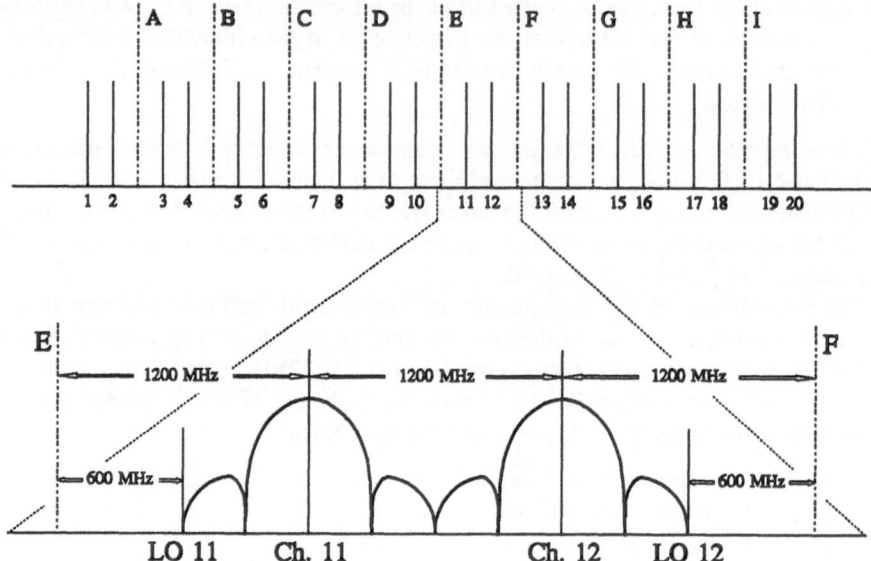

Fig. 4.3 - Channels allocation plan for UCOL

The distance of each channel with both the nearest adjacent channel and the image of the opposite one has been fixed to 1200 MHz. This spacing is a multiple (four times) of the bit rate and fits in the minimum of the interfering spectrum; moreover, this configuration allows for the maximum channel density for the same channel spacing.

The *Carrier-to-Interference ratio* (C/I) resulting from this channeling scheme together with the DPSK modulation spectrum dramatically affects the receiver sensitivity. In order to reduce ACI, data filtering before transmission will be employed (with a cut-off frequency equal to the bit rate), which determines a C/I higher than 20 dB with a penalty of 0.5 dB at a BER of 10^{-6}.

Higher channel density might be possible by both decreasing the IF to the next allowed lower value (i.e. 450 MHz) or by reducing the spacing between channels to 900 MHz. The first option would reduce the receiver sensitivity due to the residual IF signal in the baseband, caused by the partial superposition of IF and BB filters; the other would cause an ACI increasing.

4.5 Reference Carrier Generation

J. Bekooij

As it has been mentioned above, a reference frequency comb is generated by mode-locking of an external cavity semiconductor laser. This comb, consisting of a number of equidistant lines, can be generated by mode-locking of a grating extended semiconductor laser. Mode-locking is the process of phase-locking of successive longitudinal external cavity modes by gain modulation at a frequency resonant to the *Free Spectral Range* (FSR) of the cavity. The spectrum of the MLL consists of a cluster of equidistant lines (spaced exactly by the modulation frequency) in a bell-shaped envelope. The extension of the comb is limited by the grating spectral filtering.

This approach seems a promising technique in a FDM coherent optical communication system. It offers a good comb stability; in fact mode hopping does not occur in a mode-locked laser because of the existing coupling between the longitudinal modes and it is expected that a very comprehensive spectrum can be created. This implies that a FDM communication system with a frequency reference system based on a mode-locked laser has potentially a large spectral bandwidth capacity.

4.6 Carrier Generation for the Individual Channels

O. Koning

In order to derive the optical channels and the LO frequencies according to the allocation plan, the TNL is used as a frequency shifter element. Therefore, this device is able first to select the required comb line, respect to which the output frequency is properly shifted, and then lock it on by means of an AFC system. It has been already demonstrated that the AFC technique permits locking with a low input level of the reference lines allowing therefore the use of wide comb spectrum. Moreover this approach is conceptually simple and exhibits good frequency stability performance. The proposed scheme requires as a key component a TNL module continuously tunable over the entire UCOL frequency range around the nominal system wavelength. Besides, it requires a linewidth less than 1 MHz, due to the choice of both DPSK modulation format and a 300 Mb/s data rate, and an optical output power of 0 dBm. At the start of this project, semiconductor lasers fulfilling this requirement were not commercially available, For this reason the project aimed at the realisation of external cavity lasers, consisting of a normal semiconductor laser with an extended cavity to reduce the linewidth. For implementation in realistic systems, these lasers are expected to have a worse performance with respect to long term stability, tuning and robustness. For this reason, developments on the market are followed to study whether commercial devices (DBR or DFB quantum well lasers fullfilling the UCOL requirements become available. In fact, at the end of the project DBR-multi-quantum well lasers have been bought to be included in the demonstrator link, as discussed in Chapter 7.

An alternative method foresees an Er-doped laser with a grating acting as tuning element. This device is pumped by a semiconductor diode laser operating at 980 nm and has an output power in excess of 0 dBm across its tuning range. The normally broad free-running linewidth of the fiber laser is narrowed by incorporating a fiber grating as an output coupler. This grating acts as a narrowband reflector which is capable of forcing the laser to oscillate in a single longitudinal mode. The laser frequency can be tuned by either temperature tuning or stretching of the fiber grating using a piezo-electric element.

The required time for the frequency locking, which is an important parameter for the initialization procedure of the UCOL system, is mainly depending on the tuning characteristics of the TNL. Locking times of around one second seem to be feasible with both approaches; for DBR-lasers it can be much smaller if fast tuning is performed by the current adjustement of one section.

5 Network Extension

G. Veith

During recent years optical amplification has gained enormous interest as it in general provide an attractive means to upgrade the capability of fibreoptic communication systems. *Erbium-Doped Fibre Amplifiers* (EDFAs) have been developed in a revolutionary way from their first laboratory demonstrations (in 1987) to field-deployable prototype modules available today. The benefits of this advanced amplifier component have been demonstrated in impressive laboratory system experiments during recent years (e.g. high bit-rate digital long-range transmission, analogue CATV distribution). EDFAs are considered to be a key component for the near-term implementation of fibre optic systems into the subscriber area (FTTC, FTTH) and will play an important role for the introduction of future *Integrated Broadband Communication Networks* (IBCN). Due to its unique properties, the EDFA offers also attractive features for all types of optical multichannel transmission systems (WDM, HD_WDM, FDM or *Coherent Multi-Channel*, CMC). Therefore, in order to recognize the rapid progress and growing importance of this key component, some activities on EDFA and a study on its impact on multichannel system applications have been introduced into the UCOL workplan, specifically for investigating EDFA performance parameters relevant for an optional upgrading of the proposed system demonstrator.

The main features and performance parameters of EDFAs with respect to optical multichannel system applications as CMC, WDM and HD_WDM are: a) High fibre-to-fibre gain (20 - 30 dB); b) broadband gain bandwidth (typ. > 30 nm or 3.75 THz); c) high saturation output power (typ. > 15 dBm); d) low noise figure (typ. 3 - 5 dB); e) no interchannel crosstalk; f) bit rate and modulation format transparency; g) polarisation insensitivity.

EDFAs offer high effective optical fibre-to-fibre gain (20 - 30 dB) which can be efficiently used for a compensation of optical link losses, distribution losses (e.g. splitters, passive star) and coupling/insertion losses of passive optical components along the transmission path. A highly effective gain and/or output power are specifically important for CMC or HD_WDM systems, where the power budget of individual channels is substantially reduced by the multiplexing (combiner) stage of all channels onto one fibre: a combination of 2^N channels onto one fibre adds up to (Nx3) dB insertion loss per channel. One single EDFA inserted after the combiner stage could easily compensate for the combination losses of 64 signal channels (18 dB).

In general, since the EDFA offers a nearly uniform gain over a broad spectral range (typ 30 nm, corresponding to 3.75 THz bandwidth) it can enhance simultaneously the power budget of all individual signal channels of an optical multichannel system (CMC, WDM, HDWDM) without degrading the signal transmission quality. Assuming e.g. a channel bit rate of 2.5 Gb/s and a channel separation of 25 GHz the available optical gain bandwidth of an EDFA (3.75 THz) would allow for a simultaneous optical amplification of 150 channels, corresponding to an overall transmission capacity of 375 Gb/s.

The main limitation for the use of EDFAs in the above mentioned multichannel application may arise from gain / output power saturation. The maximum output powers of practical diode pumped EDFA booster modules are < 20 dBm, which limits the individual channel power level of a 100-channel system to < 0 dBm.

Polarization independence is an important and necessary property of EDFAs in multichannel applications, since the optical amplifier has to provide all signal channels with uniform gain irrespective of their polarisation states. A polarisation control for all individual channels of a multichannel system is technically very difficult to realize and would not be economically feasible.

Bit rate transparency means that EDFA provides high optical gain independently from the signal channel bit rate up to highest speeds (> 10 Gb/s). An EDFA is also transparent for analogue/digital and IM, ASK, PSK and FSK modulated signals.

No interchannel crosstalk means that optical amplification of one channel does not affect the neighbor signal channels.

Finally, the *Noise Figure* (NF) of EDFAs (specifically with 980 nm pump) is close to that of an ideal optical amplifier (NF=3 dB) which allows for an effective amplification of signal channels without substantial degradation of the *Signal-to-Noise ratio* (S/N). An EDFA preamplifier can improve the receiver sensitivity of Direct Detection (DD) receivers close to the quantum noise limit, provided a narrowband optical filter is inserted before the receiver front-end in order to suppress the ASE-ASE beat noise. Thus the receiver sensitivity of a DD receiver + EDFA preamp could be close to that of an ideal coherent heterodyne (FSK, DPSK) systems (~ 40 photons/bit).

To summarize, EDFAs offer high performances for the realization of all types of optical multichannel systems (CMC, WDM, HDWDM) in point-to-point, point-to-multipoint, multipoint-to-multipoint configurations. The implementation of EDFAs keeps the way open for a future upgrading of the system capacity if required e.g. by

- increase of the overall system transmission capacity,
- the introduction of new services or higher service quality (e.g. HDTV),
- an extension of the network-radius,
- an increase of the number of subscribers.

The potential options for an upgrading of the proposed UCOL system demonstrator have been analysed by taking into consideration the power budget requirements of the reference comb and th signal carrier distribution subnetworks.

Apart from the general benefits resulting for multichannel systems by the substantial improvements of the overall system power budget result as the main conclusions:

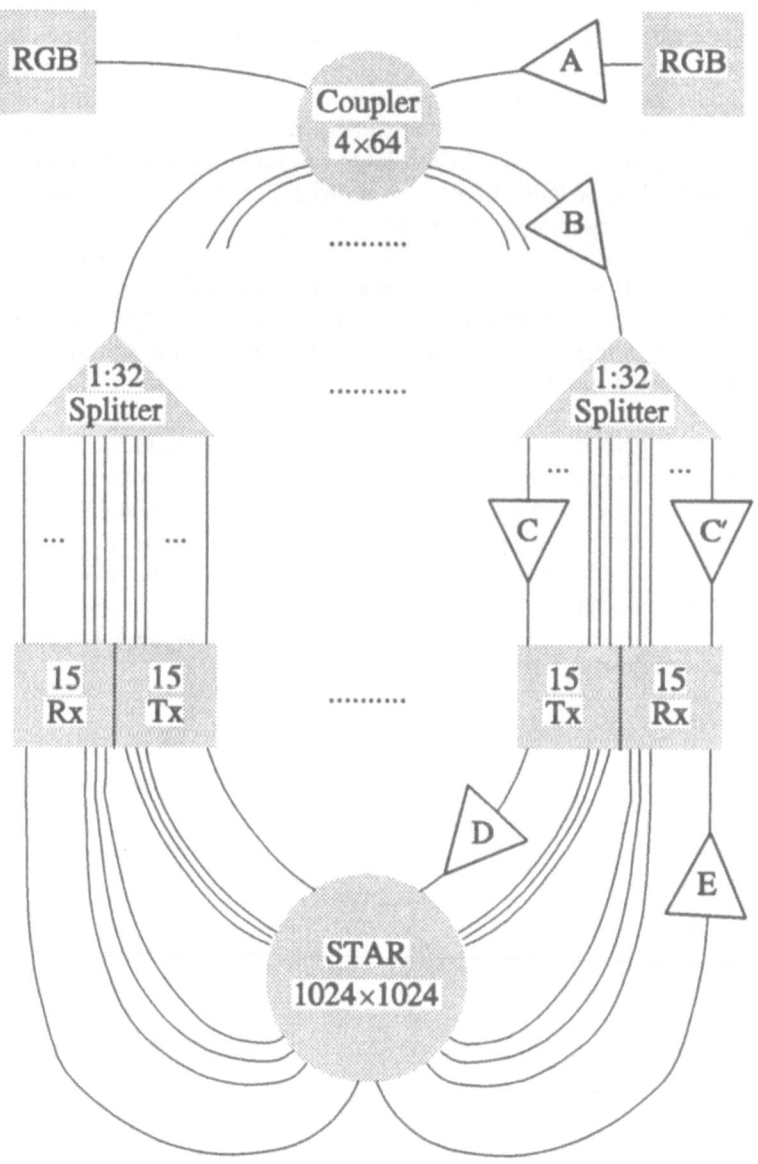

Fig. 5.1 - Options for an introduction of EDFAs within UCOL passive star network.

- EDFAs can be efficiently used for an upgrade of the UCOL overall system performances by introducing these amplifier components into the signal transmission path as well as into the reference comb distribution path (see Fig. 5.1);
- the potential number of network interfaces/substations could be considerably increased by the introduction of decentralized combiner/amplifier and amplifier/splitter stages. The use of 15 dB amplifiers and of appropriate 1:32 splitters (32:1 combiners) would allow for a potential connection of 2048 UCOL Stations instead of the 64 planned, whereby the 1024×1024 central passive star remains unchanged. A parallel extension of the reference comb distribution network is also feasible by optical amplifiers.
- The potential UCOL network radius can be extended to > 100 km, which allows principally to extend the original UCOL LAN system-concept to *Metropolitan Area Network* (MAN) or *Wide Area Network* (WAN) applications, provided the extensive network management task could be solved.
- The overall UCOL transmission capacity could be increased by an increase of the signal channel bit rate (e.g. up to 2.5 Gb/s). The gain added by one amplifier stage could be used to compensate for the lower receiver sensitivity at the higher bit rate. However, a higher bit rate multichannel system would require a new receiver design and a modified channel allocation scheme (increase of channel spacing).
- The number of signal channels and of the reference lines could be increased substantially due to the broad gain bandwidth, flat gain characteristics and low interchannel crosstalk of EDFAs. Assuming the actual UCOL channel bit rate of 300 Mb/s, a 40-channel coherent DPSK system operating within 1 nm of the nominal UCOL transmission wavelength could be realized by the introduction of EDFAs with enhanced power budget and system margins. An extension of the channel number, however, would have strong implications on the network management protocols.

6 Experimental Results: Developed Components

6.1 Multi-line Master Source: Mode-Locked Laser

O. Koning

In the UCOL network all transmitter and receiver lasers are locked relative to a comb of reference frequencies. This optical reference comb consists of a number of equidistant lines and is distributed to all stations by means of a separate optical star. The main advantage of this referencing technique is in the common frequency control at distant network locations.

The requirements on the optical comb are imposed by the system concept, and have been modified several times during the first year of the project; for practical reasons has been then decided that the UCOL multi-line master source must generate a comb consisting of 9 lines with optical power per comb line larger than -15 dB and spaced by 3.6 GHz.

During the first phase of the project the requirement on the linewidth of individual comb lines was 100+450 kHz. In this case locking of tunable lasers relative to the comb was performed by an *Optical Phase Locked Loop* (OPLL). However, from calculations, it appeared that the predicted profitable phase noise cancellation of the OPLL was too small to assure proper system performance. Therefore, frequency locking to the comb by AFC techniques was adopted, setting a less severe requirement of about 1 MHz to the linewidth of the comb lines.

Initially, it was proposed that the necessary identification of the distinct comb lines was performed by measuring the power of the individual lines. This may be performed, for example, by monitoring the beat spectrum of a scanning laser with the comb. However, since the power differences among the comb lines is rather small an alternative channel acquisition procedure was proposed. In this procedure the locking of the laser is performed by tuning the laser within the AFC locking range of the required comb line; this can be achieved by making use of laser parameter tables (i.e. relation between current and temperature, and wavelength). After this coarse tuning, the AFC-loop closes, and high stability locking is possible. Note that absolute frequency stabilization of the comb may be necessary for this approach.

An optical reference frequency comb can be generated by mode-locking of a semiconductor external cavity laser. In a mode-locked laser the gain is modulated at a frequency equal to the FSR of the cavity. In that case sidebands are created efficiently, resulting in a set of lines spaced exactly by the modulation frequency. The number of sidebands can be increased by increasing the modulation power, but is limited by intra-cavity filtering; the total laser output is distributed over all lines.

The main advantage of mode-locking of a laser over external modulation consists in avoiding insertion losses; furthermore, the power distribution of the various comb lines generated by mode-locking is relatively smooth compared to comb generation by external frequency- or phase-modulation. This gives basically the opportunity of extended comb generation. Finally, the inherent stability of the comb spectrum is large because, since the cavity modes are coupled when mode-locking, mode hopping cannot occur.

For application in the UCOL system, a grating external cavity laser has been designed. This laser consists of a Fabry-Perot laser diode, a collimating lens and a reflection grating. The diode facet facing the external cavity is AR-coated to achieve a single cavity. A comb-line spacing of 3.6 GHz is obtained for a cavity length of about 4 cm. Discontinuous tuning of the laser over a range of about 100 nm is possible by changing the angle of the grating. For continuous tuning, over about 50 GHz, the tilting of the grating is accompanied by a proper change of the cavity length.

The resolving power of the grating is determined by the number of illuminated grooves, and can be changed either by changing the grating type or the dimension of the intra-cavity beam. It is proved that any intra-cavity filtering significantly influences the comb envelope or can cause spectral instabilities. Therefore, reflections from the diode facet facing the cavity and the collimating lens had to be minimised.

The characterisation of the MLL has been focused mainly on spectral properties of the source. For a coarse (0.1 nm) determination of the laser wavelength a grating based optical spectrum analyser was used; measurement of the comb spectrum with higher resolution was obtained with a Fabry-Perot interferometer. Even higher resolution was possible by optical beating experiments; the beat spectrum of the comb with a second (high spectral purity) laser was examined by means of an electrical spectrum analyser.

Besides the generation of an optical frequency comb, experiments on locking of a tunable laser to the reference comb have been performed. The optical output of a tunable laser and the comb generator were combined on a photodiode by means of a fibre coupler; an AFC, having a 300 MHz zero-crossing, was used for frequency locking the tunable laser relative to one of the comb lines. During the tuning of the entire comb over a frequency range of about 40 GHz, locking relative to the distinct comb lines remained stable.

During the first year of UCOL the work was mainly focused on both experimental and theoretical study of the principle of comb generation by mode-locking, while in the second year the aim was to build a prototype comb generator. In this prototype a micro-strip printed circuit has been constructed for an efficient connection of both the DC bias current and the RF modulation current (2+8 GHz). Moreover, temperature stabilisation of the laser diode and provisions for the optical couplings both to the external cavity (by a collimating lens) and to the system were realised. Finally, the laser module needed stable mounting on the optical bench used to build the external cavity.

Two approaches have been used for output coupling. Firstly, a tapered fibre has been used which was fixed close to the laser diode facet; the gold coated fibre was connected to the laser module by using a selected soldering alloy, with melting point of 103 °C: however, due to the large shrinkage of the solder, the taper fixation process proved to be unreproducible. In the second and more successful approach a specially designed laser diode carrier was used, which allowed for output coupling by lenses at both laser facets; at the same time the inclusion of a bulk optical isolator was possible. This second approach proved to be highly stable and is currently used in the prototype.

Since a largely divergent beam emerges from the laser diode, a high *Numerical Aperture* (NA) collimating lens is needed; because of their small dimensions, GRIN-lenses have been used, so enabling small cavity lengths and, consequently, comb generation at large line spacings. A GRIN-lens holder, providing high resolution adjustment, was developed. Nevertheless, from experiments it appeared that reflections from the lenses output facet caused the comb envelope to become strongly undulated. Therefore, it has been decided to switch to angled facet and plano-convex GRIN-lenses, so minimizing the influence of reflections.

Several grating types with different resolution have been used during the experiments; these gratings have been placed on a mount which allows for both manual and electrical adjustment (by piezo elements). In order to minimise drift and hysteresis, piezos with built-in position sensors have been used. The external cavity is mounted on an optical bench placed in a temperature controlled and isolated laser housing. The FSR of the external cavity can be adjusted from 3 to 9 GHz, corresponding to a cavity length of 5 to 1.8 cm. The generator of the modulation frequency consists of a 2÷8 GHz low-noise synthesized signal generator with an RF power amplifier.

Presently, the prototype comb generator is able to generate a comb consisting of 9 lines, spaced by 3.6 GHz. A power per comb line of minimally -13 dBm is obtained, with a linewidth of 100÷600 kHz. Stable operation for several months has been observed, so demonstrating the applicability of the MLL in the UCOL system.

The mode-locked master laser is a free running laser, which tends to drift in frequency. As all network interfaces are locked to the comb, no channel interference during operation will occur. However, for (re)start up procedures it is advisable to have a well defined operating wavelength. This can be realized by stabilizing the wavelength of the comb generator with respect to an absolute reference. This can for example be a gas cell (with well defined spectral absorption lines), or a stable laser (e.g. a gas laser). The feasibility of this absolute referencing is shown in Fig. 7.4. Part of the comb spectrum is branched off and optically mixed with the light from a master laser. AFC-technology is used to lock the comb to the reference frequency of the master laser.

6.2 Er^{3+}-doped Tunable Narrow-line Laser

G. Cowle and L. Reekie

In the initial stages of the project, it was necessary to identify the optimum wavelength for pumping the Er^{3+}-doped fibre laser, given the need for a diode laser pump source. *Excited State Absorption* (ESA) measurements were carried out which identified λ=980 nm as being the ideal pump wavelength, since the more obvious possible absorption band at λ=807 nm suffered from severe pump ESA. However, diode lasers operating at λ=980 nm were not commercially available at that time, so Er^{3+}/Yb^{3+}-co-doped fibre lasers pumped around λ=800 nm were developed as a possible alternative source. The power specification for the TNL was met in a diode laser pumped, untuned device.

Various methods were used to narrow the operating linewidth of the Er^{3+}-doped fibre laser in order to meet the UCOL specification. The initial preferred option of incorporating a fibre grating as a narrowband feedback element in a short linear cavity was postponed due to problems with the manufacture of gratings and highly doped fibre. Instead, an Er^{3+}-doped resonant ring laser incorporating an optical isolator was used to create travelling-wave laser in order to eliminate the spatial hole burning which would oridinarily lead to multi-longitudinal mod operation. Using this technique, single longitudinal mode operation was obtained with a linewidth of $\Delta\lambda$<60 kHz.

This cavity design met the UCOL specification for wavelength, linewidth and output power, but it was not possible to tune the laser over the required wavelength range due to the lack of a suitable tuning element. This problem was overcome by adopting an anti-resonant ring structure incorporating an optical isolator and a fibre grating. Diode laser pumped operation of this laser has been achieved using commercial 980 nm lasers (Spectra Diode Labs model SDL-6300), and a linewidth of $\Delta\lambda$~10 kHz has been obtained with an output power of 10 mW.

6.3 Active Phase Modulator and Switch

A. Labrujere

Within UCOL transmitter subsystems, TNLs are followed by an external phase modulator and optical switch. The phase modulator is used for data modulation in the DPSK format at a bit rate of 300 Mb/s. The fast switch is necessary to connect the transmitter to the network only during the transmission of signal bursts, and to disconnect it during idle states, so allowing for TDMA on each coherent channel.

In the first phase of UCOL, the processing functions should be performed by integrated optical devices; for this reason, a passive phase modulator fully integrated with a high isolation On/Off switch had to be realised, using LiNbO$_3$ technology. During the project, a second promising technique to realise the processing has been adopted: as it appeared that *Semiconductor Lasers Amplifiers* (SLAs) can be applied as phase modulators and optical On/Off switches, a feasibility study to its applicability within UCOL started.

The attractivity of active phase modulation by a SLA resides in the integration of both processing functions, modulation and switching, and in the possible future monolithic integration with a transmitter laser. An even greater advantage is that the processing functions are accompanied by signal amplification, whereas passive devices (e.g. $LiNbO_3$) suffer from serious attenuation; by implementation of active phase modulators, an expansion of the UCOL system gain can hence be realised.

A SLA is basically a laser diode, which is biased below threshold. Light injected into the amplifier experiences gain if the process of stimulated emission dominates over the band gap absorption. The signal gain depends on the applied bias current, but is anyway limited since intrinsic laser action of the optical amplifier must be avoided; to achieve maximal gain the amplifier facets are therefore AR-coated.

The operation as optical On/Off switch is based on the bias current dependence of the gain. An experimental study shows that, at an operating bias current of about 60 mA, an (unsaturated) fibre-to-fibre gain of 10 dB is feasible; for a bias current of 0 mA the amplifier attenuates the signal by about 50 dB, due to the normal band gap absorption. By switching off the bias current, an isolation degree of about 60 dB is hence possible.

The switching time of the amplifier depends on the recombination time of the carriers: theoretical results shows that the switching time to realise the required isolation degree of 50 dB is smaller than 10 ns.

The operation as phase modulator is based on the relation between refraction index of the active material in the optical amplifier and the bias current; therefore, modulation of the bias current leads to modulation of the optical path of the passing light and, as a consequence, in phase modulation. A theoretical and experimental study shows that the phase modulation response is in the order of 0.1÷0.3 rad/mA for a modally biased amplifier.

The bandwidth depends on the dynamics of the carriers in the active region; at a bias current of 60 mA, the 3 dB bandwidth is in the order of 500 MHz: for applications up to the UCOL nominal bit-rate (300 Mb/s) this bandwidth is sufficient.

The optical amplifier is basically a booster amplifier; for this reason the saturation effects are important, as they limit the maximal output power. In the presently used amplifier, an in-fibre output power of -3 dBm could be obtained without seriously affecting the phase modulation response. The maximal output power can be further increased by improving the fibre-to-chip coupling and by using an intrinsically better amplifier (e.g. with an higher AR-coating quality): an in-fibre output power of more than 0 dBm is expected to be possible.

During the feasibility study it appeared that some basic processes could give significant degradations of the receiver sensitivity; in fact, the active phase modulation process is accompanied by intensity modulation. Without any adaptation in the receiver, a penalty of about 3 dB over the DPSK receiver sensitivity must be included: however, by adjusting the receiver decision threshold, this penalty can be decreased to about 1.7 dB. An even lower penalty can be obtained by making use of interferometric effects in the SLA.

A second degrading process is induced by the SLA spontaneous emission of radiation, which beats with the LO signal and gives an *Additive White Gaussian Noise* (AWGN) like contribution to the receiver. As many active phase modulators operate simultaneously, this noise contribution increases proportionally; in UCOL the penalty ranges from 0.7 up to 2.2 dB, depending on the number of available optical channels (20÷100).

The obtained results are used to make a power budget comparison between the passive (LiNbO$_3$) approach and the active one; including the described penalties and the saturation effects, it appears that, by adopting the active approach, a system gain increase of more than 7 dB is feasible.

6.4   Polarisation Diversity DPSK Receiver

The optical receivers within the UCOL NIs are based on coherent detection, in order to achieve high sensitivity and the necessary channel-selective detection. A change in the SOP of the incoming signal has no influence on the detection quality, since polarisation diversity is applied.

At the receiver, after combining with the LO field by means of a 3 dB coupler, the signals are split into two orthogonal polarisation states by a TE/TM splitter, and each branch is led to a separate detector section. Balanced detection in each branch has been adopted in order to suppress detector noise caused by RIN of LO laser; each part provides balanced detection by means of a photodiode pair. Each balanced detector is followed by a low-noise preamplifier and an additional amplifier, after which the double DPSK demodulation is performed.

The receiver has been developed in three separate parts, as shown in Fig. 4.2: the optical hybrid, composed by 3 dB coupler, TE/TM splitters and photodiodes array, the preamplification stage and the DPSK demodulator. These separately developed components have been integrated by one of the partners to form the total receiver.

6.4.1   Optical Hybrid

B. Hillerich

The fibre-based optical hybrid for the UCOL polarization diversity receiver is shown in the first block of Fig. 4.2. It consists of a 3 dB coupler, two TE/TM splitters and a photodiodes array. To achieve low losses, all couplers are made of standard *Single-Mode Fibres* (SMFs) by *Fused Biconic Taper* (FBT) coupler technology. The essential part of the hybrid is the TE/TM splitter. Its operation is based on the polarization dependence of the two lowest order modes interfering in the fused region. For the optimisation of the couplers for polarisation beam splitting a polarisation modulator has been introduced, which allows the on-line measurement of the *Degree Of Polarisation* (DOP) during the manufacturing process; in this way a polarisation extinction ratio of 23 dB at the peak wavelength

(1535 nm) has been obtained with an insertion loss of 0.3 dB: moreover, the spectral width for an extinction ratio of -15 dB is 17 nm. In order to avoid a sensitivity penalty in the polarization diversity receiver, the signal SOP has to remain unchanged between the 3 dB coupler and the TE/TM splitters. Furthermore the power of the local oscillator has to be equally distributed to all output fibres. To ensure this, in an earlier version with 3 dB coupler and TE/TM splitters in separate housings, not only the fibre connecting the local oscillator but also the fibres between the 3 dB coupler and TE/TM splitters had to be *Polarization Maintaining Fibres* (PMFs) being spliced to the standard SMFs in correct orientation. In the new integrated design, two TE/TM splitters and a 3 dB coupler are integrated in a single package, thus avoiding the PMF in between the couplers. At first, two TE/TM splitters are separately drawn; then a 3 dB coupler is made from two of the output fibres; finally the complete hybrid is packaged in a common housing.

With balanced receivers front-ends, as used in the UCOL receiver, a satisfactory suppression of the LO excess noise (RIN) requires well matched photodiodes pairs, which have been realised by monolithic integration on semi-insulating InP substrate. The coupling of the fibre pigtails to the monolithically integrated photodiode pairs is performed by a flat pre-aligned coupling technique: to obtain this, a small *Printed Circuit Board* (PCB) has been developed, on which silicon

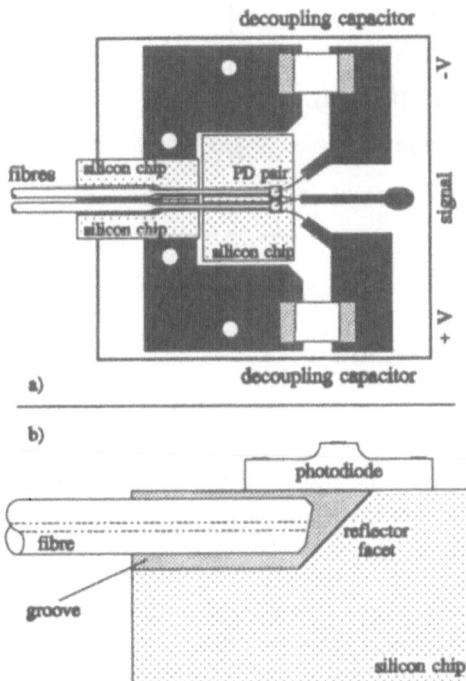

Fig. 6.1 - a) Layout of the fibre/photodiode coupling hybrid; b) Principle of coupling.

chips for the fibre-photodiode coupling and two decoupling capacitors for the bias voltages of the photodiodes are mounted, as shown in Fig. 6.1a. The central silicon chip contain two etched alignment grooves matched to the outer diameter of the fibres. Figure 6.1b shows the fibre-photodiode coupling arrangement; owing to angled end-faces of the fibres, back-reflection to the local oscillator is as low as -55 dB. This configuration also allow for an optimisation of the fibre-photodiode coupling losses in order to compensate the overall unbalance of the optical hybrid. Table 6.1 summarises the optical hybrid performance.

Table 6.1 - Performance of the optical hybrid.

Excess loss (signal fibre)	< 0.3 dB (@ 1535 nm)
Extinction ratio	< -19 dB (@ 1535 nm) < -17 dB (between 1530 and 1540 nm)
LO power unbalance among output fibres	< 7% (@ 1535 nm)
unbalance of photo-currents	< 1% (after fibre/PD alignment)
Photo-diodes responsivity	> 0.9 A/W
Optical path length alignment	within 3 mm
LO input fibre	PANDA PMF
Return loss	> 50 dB (typ.)
Total capacitance (PD pair)	1.05 pF (typ.)
PD dark current	< 20 nA

6.4.2 Low-Noise Preamplifier

E. Drijver and P. Prinz

An excellent electrical front-end should have the following properties: large bandwidth, low-noise contribution and large signal handling capability. In practice, it is hard to build a front-end having simultaneously all these properties; conventional front-ends can only score well on at most two of the three properties, and are based on one or more of the following fundamental types: *low-impedance*, *high-impedance* or *transimpedence* front-ends.

To overcome the disadvantage of each fundamental type, a new type of receiver front-end has been developed; this front-end uses capacitive current-current feedback. The feedback network ensures large signal behavior, while the use of capacitors avoids a resistive noisy network and can be made frequency independent.

To realise a low-noise current-current feedback amplifier for application in UCOL, an environment for characterisation and design had to be developed. The basic part of the environment is formed by a Network Analyser which is able to characterise electrical components. For coupling with data processing equipment special software had to be developed.

In order to realise a preamplifier on a PCB it was necessary to make a calibration set and test fixtures specially for use with PCBs. A new type of white noise source, based on self-homodyne mixing for absolute noise measurements has also been developed; a noise model has been made to predict noise contributions and select input *Field-Effect Transistors* (FETs): therefore a new FET has been bought to achieve the UCOL specification on equivalent input noise current.

In the beginning of the project a third-order (three stages) amplifier has been designed to obtain high gain with the (at that time) required bandwidth of 400 MHz, and a special PCB was designed for this front-end. In the meanwhile, however, the specifications have been tightened and a bandwidth up to 900 MHz had to be realised. As a consequence, in order not to run into severe stability problems, the number of stages has been reduced to two. From modeling it turned out that it was not possible to obtain the desired gain and bandwidth with a second-order feedback preamplifier only; therefore an additional drop-in amplifier has been added to achieve gain specification. With this modification, a front-end with second-order current-current feedback and 900 MHz bandwidth has been realised, fulfilling all UCOL requirements. For this new version, it was necessary to design a new PCB, to combine the front-end and the chosen additional amplifier.

In the first phase of UCOL project, characterisation measurements have been performed on single photodiodes from another partner and agreement has been achieved on the interface between photodiodes and front-end; these single diodes lead to the construction of balanced pairs to be used in combination with the preamplifier. The combination performed well, and the results obtained are reported in Table 6.2.

During the UCOL extension phase further improvement of the preamplifier characteristics has been accomplished. By applying noise tuning techniques the noise characteristics of the preamplifier have been improved, while preserving the specifications on bandwidth, gain, and ripple. Special attention has been paid to the influence of tuning on feedback, since incautious application of tuning can cause distortion or instable behaviour of the preamplifier. A front-end has been realized with a noise power improvement of 3.7 dB compared to the non-tuned version, however, while preserving specifications on bandwidth, ripple, gain etc. This improved noise performance corresponds to an equivalent noise current of about 3.5 pA/$\sqrt{\text{Hz}}$.

Table 6.2 - Low-noise preamplifier: comparison between target and obtained results

	TARGET	RESULTS
Bandwidth	300 - 900 MHz	0 - 1100 MHz
Ripple	± 1 dB	0.5 dB
Gain	46 dB	49 dB
Third order intercept point	> 10 dBm	> 10 dBm
Output impedance	50 Ω	nominal 50 Ω
Return loss	> 20 dB	> 20 dB
Equivalent noise current	5 pA/√Hz (@ 600 MHz)	< 5 pA/√Hz
Balancing error	10 %	< 1 %

6.4.3 DPSK Demodulator

S. Forcesi and E. Neri

DPSK demodulation can be achieved without recovering carrier reference, but using a delayed copy of the same IF signal as carrier reference, provided that delay is equal to one bit time. In this way we use carrier phase of the preceding bit as phase reference for the actual bit. In a generic DPSK demodulator, the modulated carrier is filtered by a bandpass filter centered on the IF and amplified; a multiplier combines the signal with a delayed copy of the signal and, finally, a lowpass filter is required to reject the high frequency components at the multiplier output. Besides these basic blocks, the following additional circuitry for digital signal regeneration is required: a sampler-regenerator circuit and a clock recovery system for bit synchronization. This scheme is heavily affected by three main requirements of the UCOL receiver, which have an important influence on the demodulator structure:

1) insensitivity to the incoming light signal SOP;
2) large power variation of the incoming light signal;
3) TDMA capability.

These three features of the UCOL receiver have a key role also in determining the structure of the DPSK demodulator:

- SOP insensitivity in the UCOL receiver is realized using a polarization diversity scheme: vertical and horizontal polarization components are separately detected and corresponding electrical signals have to be separately processed. A "double-arm" DPSK demodulator, composed by two identical basic demodulators, is then required. Electrical power can be assumed randomly distributed in the two arms of the demodulator, so every arm has to handle signals with a wide dynamic range.

- Large power variations of the incoming light signal imply large variations of electrical power at the demodulator input. An AGC circuit, operating on the total electrical power in the two arms of the demodulator, is then required.
- TDMA requires the receiver to be able to demodulate nearly instantaneously small bursts coming from different transmitters and with different power. This ability calls for a very fast AGC and for a *Clock Recovery Circuit* (CRC) capable to recover bit synchronization in a limited time interval.

The development of the UCOL DPSK demodulator has been divided in four steps:

a) design and realization of a 100 Mb/s double arm demodulator (without AGC and signal regeneration) and corresponding test jig for continuous mode transmission (DPSK encoder and noise generator): this step was useful to gain experience on DPSK modulation and to verify some previously simulated results.

b) development of the basic circuitry of a 300 Mb/s demodulator (IF and baseband amplifiers, IF and baseband filters, delay line multiplier, signal regenerator). Using these basic buildings blocks a single-arm demodulator prototype (without AGC and with an external reference clock) was realized. At the end of this step, the performances of this simple demodulator (operating in continuous mode) were measured using a 300 Mb/s optical front-end simulator. Moreover the availability of a complete single-arm modulator/demodulator chain was very useful in the clock recovery circuit development.

c) design and realization of fast CRC and AGC for TDMA mode demodulation.

d) realization of the final double-arm demodulator (including AGC and CRC), working in TDMA mode.

IF Filter. Main requirements for the IF filter are: a) an optimum bandwidth equal to 2xBitRate, i.e. a 300-900 MHz bandpass; b) a flat time delay in the bandpass. A microstrip filter on alumina substrate has been designed using TOUCHSTONE simulation to match these requirements; this filter has been realized and exhibits the required characteristics with 0.7 dB of transmission loss.

Baseband Filter. Simulation results indicated a fifth-order Bessel-type filter with a 195 MHz bandwidth as the optimum filter at the multiplier output. Three Bessel-type filters with different bandwidth (180 MHz, 195 MHz and 210 MHz) were therefore realized and tested; best performances were obtained with the 195 MHz filter, according to simulation.

IF and Baseband Amplifiers. An IF amplifier board with variable gain was realized using a two stage configuration. The first stage is constituted by a 13 dB gain amplifier; the second stage was realized using a GaAsFET device for gain regulation (minimum gain 2 dB, maximum gain 7 dB). The baseband amplification boards has been realized in a single stage. The performances of the IF amplification board are summarized below:

- gain variability (@ 300-900 MHz)	15 + 20 dB
- gain flatness (@ 300-900 MHz)	± 0.5 dB
- S_{11} (@ 300-900 MHz)	< -17 dB
- S_{22} (@ 300-900 MHz)	< -13.5 dB
- output power (@ 1 dB gain compression)	> 12 dBm

Delay Line Multipliers. Signal is splitted by a resistive power splitter, and part of it is delayed by one bit-time using a cable of appropriate length; the two parts of the signal are then fed to the multiplier. The main requirement of this multiplier is the "linearity range", i.e. the capability of multiplying signals without compression of output power. Ideal multipliers show 2 dB output variation for 1 dB variation of both input signals. Due to the limited linearity range of passive double balanced mixers, which produces a penalty on demodulator performances, an active mixer (Avantek IAM 82018) has been selected and tested, showing wide linearity range, a good input and output matching (thus eliminating problems due to reflections of the delayed signal) and a low input power.

Signal Sampler and Regenerator (or Decision Circuit). For data regeneration demodulated signal has to be sampled at the instant of maximum eye aperture, and compared with a threshold at the same distance between the two levels of the eye-diagram; such a regeneration system is composed of a comparator and a master-slave flip-flop.

Automatic Gain Control Circuit. In UCOL the same optical channel is time shared between different transmitters; for this reason the optical power of each burst at the receiver input can assume any value in a wide range, depending on different attenuations of the optical path. A very fast automatic gain control (AGC) circuit is then necessary to assure that the level of the demodulated signal at the decision circuit input (sampler-regenerator circuit) does not depend on optical power and is equal for all bursts. An AGC circuit directly stabilizing the level of the demodulated signal is difficult to realize because the variability range of the delay-line multiplier output level is double than the input one. Therefore the AGC circuit has to stabilize the total IF power, which is randomly distributed on the two demodulator arms depending on the optical SOP; it is important to underline that the AGC circuit has to control the sum of the powers on the two channels without changing their ratio. Main requirements of the AGC circuit are:

- capability to control input signals within 15 dB of power range;
- total transfer function with flat magnitude and linear argument in the bandpass of the IF signal (300-900 MHz);
- transient time at the beginning of each burst as short as possible (less than 200 ns);
- less than 1 dB of output power range for any input level and power distribution on the two channels;
- less than 1 dB of difference between AGC input and output power ratios on the two channels.

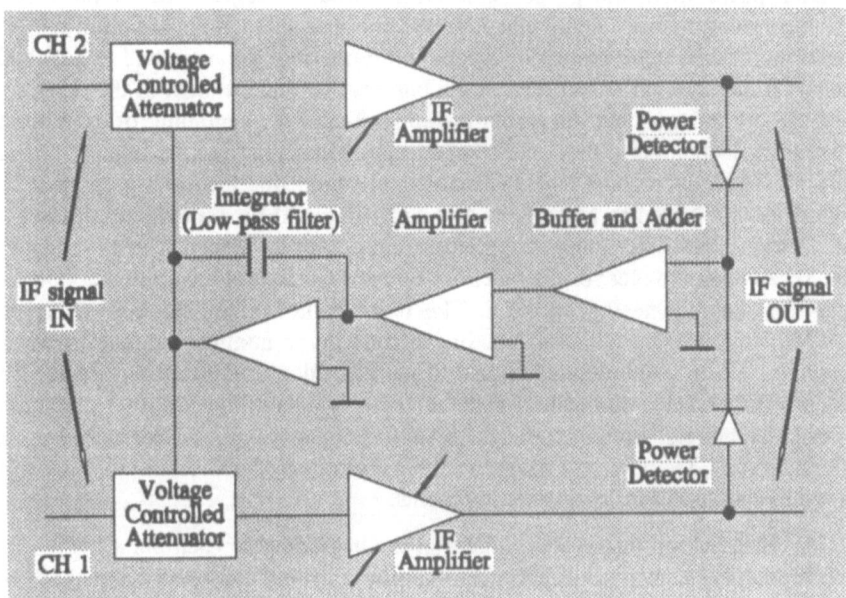

Fig. 6.2 - Automatic Gain Control circuit

A suitable scheme for the AGC is shown in Fig. 6.2. This scheme is based on a feedback control loop: signal power is separately detected on the two channels and the outputs are added, amplified and filtered to drive a voltage controlled attenuator. This device plays a key role: its main characteristics should be a very fast response to step variations of the control voltage and a flat attenuation in the 300-900 MHz band. A MMIC GaAsFET attenuator chip (HP HMMC 1001) with sub-nanosecond response and very high bandwidth has been chosen; in this device the attenuation presents a logarithmic dependence on the control voltage. The prototype AGC circuit has been assembled and tested: it fulfills the initial requirements with a transient time at the beginning of each burst of about 60 ns and a gain flatness lower than ±0.5 dB.

Clock Recovery Circuit. For signal regeneration it is necessary to recover a bit synchronization signal (or "clock") as a time reference for the sampler-regenerator circuit; an ideal time reference has equally spaced zero crossings, perfectly in phase with optimum sampling time. Non-ideal recovered clock is affected by two main impairments:

a) phase jitter - recovered clock is a noisy reference, i.e. its phase is randomly distributed around a mean value (phase jitter is usually characterized by the mean squared value of the phase deviation);

b) phase static error (or phase offset), i.e. a phase difference between the ideal time reference and the mean value of the recovered time reference.

Both phase jitter and phase static error introduce degradation of the receiver performances, because demodulated signal is not sampled at the optimum sampling time. In TDMA communication systems information is transmitted in successive bursts from different transmitters; each burst is then independent of the others, and requires an independent bit synchronization; i.e., at the beginning of each burst is necessary to recover a new clock, with a particular phase and a slightly variable bit-rate. Each burst consists of two parts: a preamble and the message portion. The preamble is divided in two sections: a signal interval for bit synchronization and the "unique word" for burst synchronization; during the first part of the preamble, the receiver has to recover the clock for correctly sampling bits in the unique word and then in the message portion. The basic approach for clock recovery in a TDMA environment with NRZ coding is based on a nonlinear signal processing: received IF or demodulated signal is processed by a nonlinear device (e.g. square-law rectifier, or absolute-value rectifier, or delay-line multiplier) to generate a discrete spectral component at the bit-rate frequency, which is extracted using a "tank circuit", i.e. a narrow passband filter. Moreover, the CRC in UCOL receiver have to satisfy the following requirements:

1) fast acquisition of the clock phase at the beginning of each burst;
2) capability to handle any power distribution between the two demodulator channels, due to variations of the incoming signals SOP and total optical power;
3) minimization of the phase jitter due to both thermal noise and signal statistic;
4) supply to the decision circuit of a constant power level (thus avoiding possible fluctuations of the sampling time).

From preliminary measurements there was some evidence that better performances can be achieved by processing the IF signal rather than baseband (demodulated) signal and by using a delay-line multiplier (with half-bit delay time) rather than other non-linear devices. Therefore the CRC chosen for UCOL receivers consists of these three basic blocks:

1) a delay-line multiplier (with half-bit delay time), fed by the IF signal, which creates a spectral line at the clock frequency;
2) an high-Q single resonant passband filter selecting this line;
3) a limiter, which avoids power level variations of the recovered clock.

The CRC has been therefore built in its final double-arm configuration, where the IF signal is derived after the AGC circuit. In this case the overall clock recovery time is roughly composed by the sum of AGC and CRC delays; simulations give a worst case delay of 32 bits, corresponding to about 100 ns. The CRC has been also tested in continuous mode operation: the sensitivity degradation introduced is negligible (0.2+0.3 dB).

A complete double-arm DPSK demodulator, including a signal regeneration circuit, fulfilling all UCOL requirements has been realized by integrating in few boards the basic circuitry of the demodulator, the final AGC board and the clock recovery circuit in its final configuration. During the realization of this final prototype great attention was paid to the matching of the two channels corresponding to TE and TM polarization; actually, perfect matching of electrical

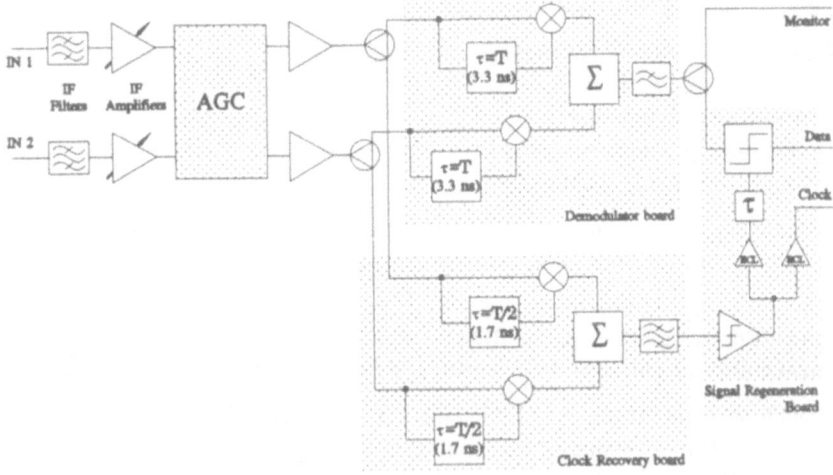

Fig. 6.3 - Complete UCOL demodulator: schematic diagram

delay and power gain of the two demodulator channels is necessary to remove BER degradations due to the polarization diversity scheme. For the same reason perfect matching of the two channels of the clock recovery circuit is also necessary. In Fig. 6.3 the schematic diagram of the final prototype is shown.

6.5 Erbium-Doped Fibre Amplifier

G. Veith

An extensive analysis of the EDFA performance parameters and a characterization and test of subcomponents and hardware configuration have been undertaken in order to get reliable data for the EDFA module optimization and for the specification of parameters relevant for system applications. Initially, these investigations have been made with not yet optimized components and devices available at that time. The considerable technological progress made recently in the development and availability of EDFA subcomponents (e.g. Er-doped fibres, pump laser diodes, fibre pump multiplexers) is also reflected by the improvements of experimental results and by the advancement of hardware configurations realized and tested in laboratory. These investigations offer a solid basis for the specification and optimization of the main EDFA performance parameters with respect to specific system applications.

The objectives of this activity were the test and the optimization of the key components and configurations necessary for the realization of an EDFA module, including:

- Erbium-doped fibre characterization and test,
- EDFA laser diode pump source,
- EDFA fibre pump couplers,
- Doped fibre splicing technology,
- Module / packaging technology.

Preliminary investigations on EDFA pumping schemes have confirmed that 980 nm is the most favorable pump wavelength for the EDFA due to:

a) highest available pump efficiency (typ. 5-10 dB/mW);
b) lowest noise figure (typ. 3-4 dB);
c) availability of polarization insensitive 980/1550 nm fibre pump couplers.

The main obstacle for the use of this most efficient pump band has been for a long time the insufficient status of 980 nm laser diode pump sources. In recent years 1480 nm laser diodes have been preferred as EDFA pump source since their technology was more advanced and since they could deliver higher output power. During this year the situation has changed: there has been a dramatical progress in the development of strained layer narrow-stripe InGaAs/GaAs chips, and 980 nm laser diode modules with high output power and good reliability are standardly offered by some manufacturers (e.g. Lasertron).

Recent tests of these new generation 980 nm laser diodes in an EDFA environment have demonstrated that they can fulfill easily the requirements derived from our EDFA analysis, including the high output power (>40 mW in fibre pigtail), the emission wavelength (980±5 nm), the threshold current (< 30 mA) and the operating current (< 160 mA).

In practical applications the pump power available in the active fibre is an important parameter for the EDFA performances and has to be controlled during the operation of the amplifier. It has been shown, that the free output port of the pump WDM fibre coupler can be used as monitor path for the pump power level provided an excitation of higher order modes is avoided in the fibre pigtails or in the fused fibre coupler region: this can be achieved e.g. by using fibre pigtails with a cutoff wavelength < 950 nm. The actual pump power can be controlled by a P/I circuit using the monitor pump power as control parameter.

With these type of pump sources the following performances can be achieved in an EDFA by using standard quality erbium doped fibres and single laser diode co-propagating or counter-propagating pump schemes:

small signal gain: ≥ 25 dB
max. output power: ≥ 10 dBm
max. pump efficiency: > 4 dB/mW

A detailed analysis and characterization of the erbium doped fibre properties have been undertaken in order to derive design parameters for EDFA modelling and to optimize the performances of the amplifier fibre for an implementation into an EDFA module. There has been a great effort during recent years to accurately model the gain of EDFAs. One commonly used conventional approach, proposed by Saleh et al. [3], derives the coupled differential equations for pump and signal wave evolution along the active fibre using the population of the associated erbium ion levels given by the rate equations. This approach has been extended by introducing the mode field profiles and the erbium distribution into the formulas giving closed form analytical relations between pump and signal input powers, fibre length and fibre gain. This modified theory allows e.g. to predict maximum gain and optimized fibre length for different sets of signal and pump input power in excellent agreement with measured values. As input parameters for this analysis are used only the nonlinear absorption characteristics of the signal and pump wave, as measured in a standard laboratory setup.

For the realization of a compact EDFA module the different subcomponents (erbium doped fibre, pump laser diode) have to be coupled with good efficiency.

Table 6.3 - Features of EDFA Lab Model

Fibre-to-fibre gain	≥ 25 dB (small input signal < -30 dBm) ≥ 10 dB (large input signal < 0 dBm)
Spectral gain region	1530 + 1560 nm
Max. output power	≥ 10 dBm (single LD pump)
Polarisation sensitivity	< 0.2 dB
Input/output port	standard SMF pigtails (9/125 um) standard SMF connectors (> 35 dB isolation)
EDFA housing dimensions	25 cm × 8 cm × 10 cm
Pump coupling element	980/1550 nm fused fibre coupler Insertion loss: < 0.5 dB (signal & pump) Crosstalk: < 20 dB (signal/pump) Polarisation sensitivity: < 0.1 dB
Erbium-doped fibre	Mode field diameter: 5 + 6 um (@ 1550 nm) Core diameter: 2 + 3 um (typ.) Erbium concentration: 30 + 60 ppm Co-dopant: Alluminium Fibre length: 20 + 50 m (typ.)
Pump laser diode	Type: Lasertron QLM9S450 Wavelength: 970 + 985 nm Fibre coupled power: > 40 mW Drive current: 150 mA (typ.) Spectral shift: 0.5 nm/°C Threshold current: 15 mA (typ.) Operating temperature: 10 + 30 °C Package: 14-pin DIL Thermoelectric cooler

For the effective transport of 980 nm pump power into the erbium fibre, 980/1550 nm wavelength selective fibre couplers have been fabricated with the following characteristics:

a) low excess loss for signal/pump (< 0.5 dB);
b) low polarization dependence (< 0.1 dB);
c) < 950 nm cutoff wavelength.

By applying specific fibre coupler fabrication techniques (twisting after tapering), 1480/1550 nm fused fibre pump couplers with low polarization sensitivity (< 0.2 dB) have been also realized.

A key problem consists also in the splicing of fibres with dissimilar mode field diameters (e.g. at a signal wavelength of 1550 nm erbium doped fibre exhibit typical mode field diameters of 5-6 um to compare with 10-11 um of standard SMFs). It has been shown that by tapering after splicing, the insertion loss of the tapered splice can be kept < 1.0 dB for the signal and pump wave (1550 nm / 980 nm). By additionally applying a thermal diffusion technique for matching the fibre, mode fields splice losses of < 0.5 dB can be achieved.

After the detailed analysis of the EDFA performances and the careful selection of subcomponents and hardware configuration, a compact EDFA module (EDFA lab model) has been realized: the main features of this lab model are summarised in Table 6.3.

6.6 Tunable Optical Filter

H. Schmuck

In the original UCOL system concept a narrowband tunable optical filter was foreseen in order to select the global reference line out of a comb of DPSK modulated carriers for a master block synchronization. Within the UCOL project, a novel SMF filter type, based on a Fabry-Perot interferometer, has been developed, which fulfills the requirements of the above specified task, and offers a number of other additional applications.

The filter consists of two fibre-pigtailed GRIN-lenses arranged to form out an open Fabry-Perot resonator (GFP filter). The free inner surfaces are reflection coated to realize a high wavelength selectivity (Finesse). The resonator length can be changed by the use of a piezo-electric transducer corresponding to a filter transmission wavelength tuning.

Extensive feasibility tests and performance analysis have demonstrated the applicability of this filter over a widely variable *Free Spectral Range* (FSR) with low insertion loss and high selectivity. The main features of the GFP-filter are:

- Widely variable FSR: 1 GHz - 50 THz
- High Finesse: > 100
- Low throughput loss: < 5 dB (for FSR > 40 GHz, F = 100)

- Single-mode fibre compatible
- Polarization independent

Two different lab models of a GFP-filter have been designed, built up and characterized. In comparison to the first device the advanced lab model presents a more compact and more stable mechanical set-up. Table 6.4 summarizes the performance and features of this filter prototype according to the original UCOL-requirements.

In an experiment, the stabilization of the GFP-filter with respect to a given frequency line of a narrowband reference laser light source has been demonstrated. In this case the filter transmission peak has been locked to the emission line of an external cavity laser (linewidth: 100 kHz) by use of an active stabilization circuitry. A dithering technique has been employed to generate control signals for the frequency stabilization. The frequency stability of the GFP-filter was better than 0.002 x FSR. Furthermore, the GFP-filter has been applied successfully as a tuning element in an Er^{3+}-doped fibre ring laser. This laser type offers promising features as tunable narrow linewidth laser source enabling a laser emission in the 1550 nm wavelength region. The ring laser can be tuned over a wavelength range of more than 45 nm (1524÷1569 nm) by use of the GFP-filter (FSR=47 nm, filter bandwidth: 0.78 nm) as an intracavity wavelength selective element. The laser output power is -10 dBm with power variation of 0.5 dB (1529÷1565 nm). Other potential application areas of the filter are:

- Fibre-optic multichannel WDM/FDM systems
- Noise suppression in optical amplifiers
- Spectral analysis instrumentation

Table 6.4 - Characteristics of the tunable optical filter.

Optical performance parameters	
Spectral operation range	1500 + 1600 nm
Continuous tuning range	100 GHz
Optical filter bandwidth	< 1 GHz
Overall throughput loss	< 5 dB
Polarisation sensitivity	No
Module features	
Housing dimensions	90 mm × 70 mm × 33 mm
Resonator length	1.5 mm
Temperature drift of absolute frequency	< 0.04 FSR/°K
Operation/control voltage	6.4 VDC/FSR

6.7 Traffic Generator Detector

A. Almeida, N. Almeida, A. Alves, P. Assis, J. Cabral, E. Carrapatoso, E. Ferreira and M. Ricardo

The original workplan of UCOL project was targeted to the development of a system capable of assessing the performance of the UCOL network by means of packet traffic that should be generated and detected. This system, called the *Traffic Generator and Detector System* (TGDS), was a real time distributed system consisting of one *WorkStation* (WS), a variable number of *Processing Units* (PUs) and one LAN to interconnect them.

The TGDS working principle was the measurements of delays experienced by packets on their way from the generator (located in one PU) to the detector (located in the same or another PU) across the network. These results should be gathered during experiments defined by the TGDS user at the graphical interface offered by the WS and shown also graphically at the WS.

Due to a revision of the UCOL workplan, it was decided to simplify the architecture so that with less investment it would still be possible to end up with a valuable and flexible device which could be used in the evaluation and test of networks. This simplified version would consist of only one PU which would interface with the user by means of an RS 232 terminal. Therefore, the functional capacities for the generation and detection should all be available but the graphical interface and the control at the WS would not be implemented.

To achieve these goals, two boards, one for the generator and the other for the detector, were designed, built and tested, based on the TMS320c25 digital signal processor. Basically, they are 16 bit boards, double Eurocard, with VME interface and a piggy-back interface to the network. The two dedicated boards are controlled by a commercially available single board computer featuring a VME interface and running the OS-9 operating system. All the software to define experiments and analyse results, to control the dedicated boards and to generate and detect packets was also written and tested. Finally, the software and hardware were put together and some experiments carried out; one conclusion that may be drawn from those experiments is that the requirements that had been identified at the beginning· of the project have been exceeded by far.

6.7.1 TGDS Architecture

As depicted in Fig. 6.4, the TGDS architecture comprises the following elements: one workstation, a variable number of processing units and one LAN (Ethernet). The workstation is the entity through which the user accesses the system. The processing units are the basic entities responsible for the generation and detection processes. The Ethernet is the communication medium between the WS and the PUs.

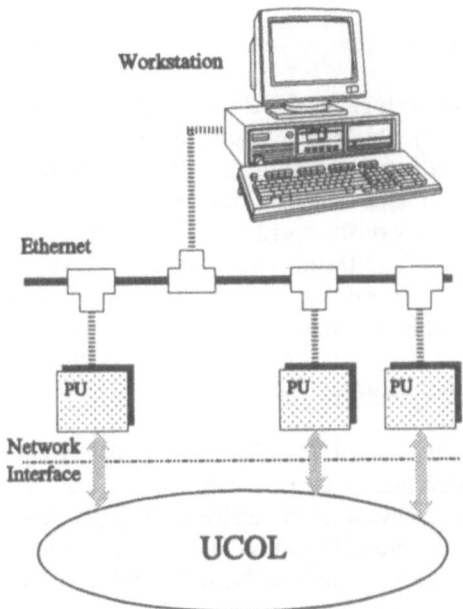

Fig. 6.4 - General TGDS architecture

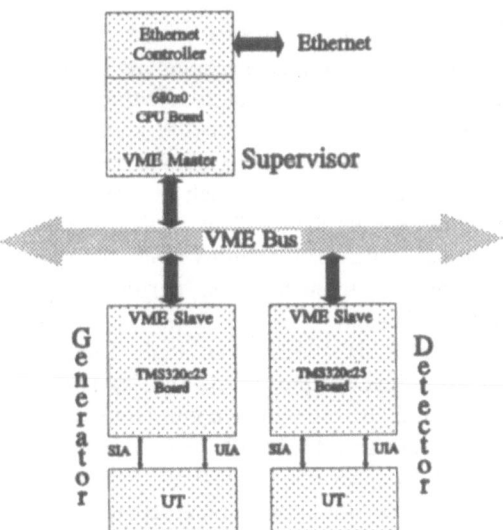

Fig. 6.5 - Processing Unit architecture

6.7.2 Hardware development

Processing Node Architecture. Each PU has three boards interconnected by means of a VME bus (Fig. 6.5). One of the boards, the supervisor, is a general purpose microcomputer. It communicates with the WS by means of an Ethernet and controls the generation and detection processes, which are executed in two dedicated boards. The supervisor board is a VME Master, based on the Motorola 68030 CPU, and supports an Ethernet controller as a piggyback. The two dedicated boards are based on the Texas Instruments TMS320c25 digital signal processor and, from the VME point of view, are slaves. They would be attached to the UCOL. The VME bus is used for the exchange of control and data between the supervisor and the other boards.

Generator and Detector Board. In order to render the TGDS even more flexible, the generation/detection boards have exactly the same architecture: any of them may be used either to transmit or to receive data. They consist of the following blocks: Program/Data Memory, Decode & Wait States Generation, TIME (real time clock), TIMER (interval generator), MAILBOX, FIFO and CONTROL/STATUS registers. The generation/detector boards are based on a single PCB, double Eurocard format. As they will be plugged in a VME backplane they must be in accordance with the mechanical and electrical characteristics specified by the latest version of the VME specification (IEEE 1014-87).

In Fig. 6.6 a functional description of the PCB is made: all the blocks which compose the generation/detection boards may be seen.

UCOL Interface. The TGDS main purpose is the assessment of the UCOL network performance by means of the measurement of delays experienced by packets traveling through the network. To carry out these measurements, packets have to be passed from the TGDS to the network and then back to the TGDS. These activities imply the need of interfacing between the two systems.

The access method considered to connect each PU to the station was via a User Access Unit, as any other external device. This choice was dictated by two main reasons: flexibility and performance. Flexibility since the system might be more easily adapted to the user requirements and performance as more complete measurements might be achieved.

Figure 3.4 presents the UAU and its two main functional blocks; UCOL Termination (UT) and Terminal Adapter (TA). The TA is assumed to be service dependent. On the other hand, the UT will be common to all UAUs. The UT could be used to interface generator/detector boards to the UAU according to the User Information Access (UIA) and the Signaling Information Access (SIA) specifications. A TA block must be implemented according to the TGDS specific functions and the UIA and SIA specifications. In fact the TGDS itself would perform the TA functionality.

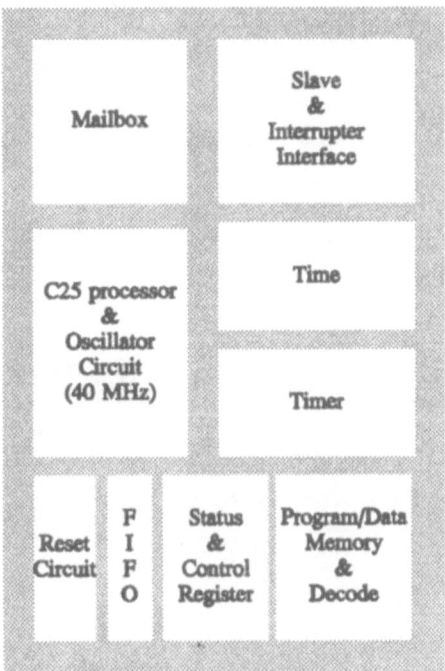

Fig. 6.6 - Architecture of the Generation/Detection board

6.7.3 Software Developments

To implement the TGDS concept a structured approach for the software was chosen. Two types of structures were considered: vertical and horizontal.

The vertical structure of the TGDS software is based on layers. The services offered by the first layer are the generation and detection of packets. The second layer, by integrating these services, offers link services: it creates an association between generation and detection. The third one offers session services: an association of links is made. The fourth layer, by using sessions, executes sessions groups or executes sessions in a specific way. The fifth layer is a graphical interface between the user and the system.

The horizontal structure is used in a layer, to divide software into smaller blocks. Blocks are independent and grouped by the proximity of functions or services they offer or by physical association. The TGDS software structure, assuming three PUs, is presented in Fig. 6.7.

Software for the WS. The software presented in this section has not been fully implemented, but was an important part of the project. The Workstation to be used was a Sun running the SunOS, which is a UNIX operating system that integrates functionalities from both BSD and System V. The SunOS has a proprietary

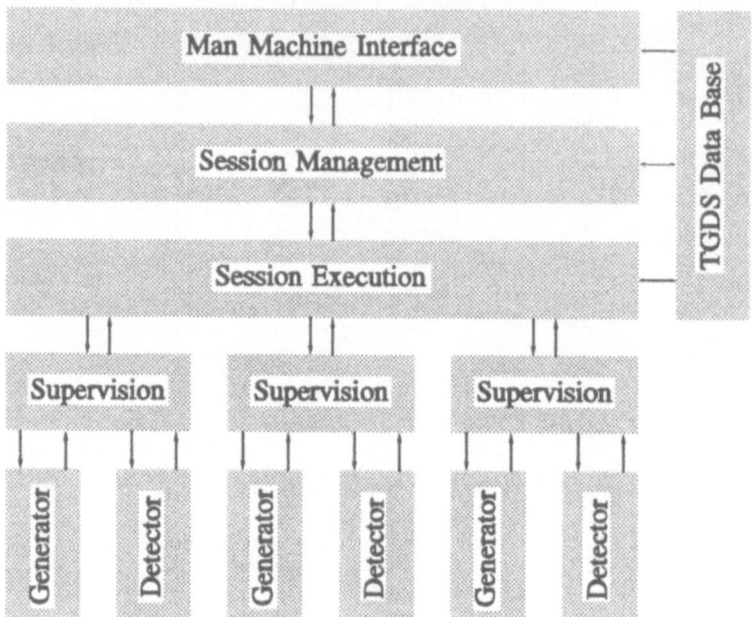

Fig. 6.7 - Software hierarchical model

graphical interface, the SunView, which offers high level functions to develop the MMI for the TGDS.

The TGDS tasks that should run in the WS were the Man-to-Machine Interface (MMI), the *Session Management* (SM) and the *Session Execution* (SE). To implement these concepts two processes and some OS facilities for interprocess communication could be used. The two processes communicate by using two sockets and one message queue as depicted in Fig. 6.8.

The message queue, that should be used to implement the session queue, supports one type of messages: session. Sockets are used to exchange messages between the two processes in order to guarantee a reliable, sequenced and unduplicated delivery of messages. The MMI should enable the user to define sessions, order their execution, and analyse results. When the user defines one or more sessions (graphical facilities should be provided to edit sessions) he can put the sessions in the sessions queue and order the start of the SE. The messages received by the SE allow the MMI to manage the queue that should be visualised in the MMI. The SE process, when given the appropriate command by the MMI, starts running the sessions queued. The SE, that runs one session at a time, will execute a session by running some functions remotely, i.e. in the PUs. The PUs will act as generator/detector servers. The remote functions were implemented using TCP/UDP facilities over the IP protocol.

Software for the PU. The PU, as we saw above, can be understood as a generation/detection server. By executing the remote commands, it can prepare

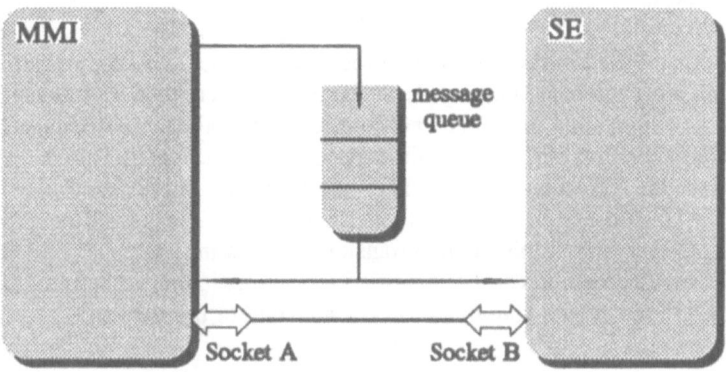

Fig. 6.8 - MMI / SE interface

data, generate according to pre-defined rules, detect and carry out measurements. The operating system chosen for the PUs was the OS-9. It is a UNIX like operating system that is ROMable and with real time characteristics. Its modular architecture enables the addition of new software modules. The TGDS tasks running in the PU are the Supervision, the Generator and the Detector. The Supervision is the PU network interface. The generator tasks generate packet traffic and perform measurements related to the generation process. The functions of the detector task are to receive the packets generated and perform the measurements.

The software actually implemented and tested is constituted by: a functions library for the C25 board, the generation, the detection, the interface functions, the server implementing the remote functions and a basic non-graphic local human interface. The functions library for the C25 boards includes the very basic device drivers for the hardware devices and the implementation of some virtual devices. The generator and detector tasks were specially developed in assembly language. The server, that is implemented by the supervision task, handles all the communications aspects of the RPCs (data transfer, timeouts, use of sockets, etc.). The application to use these functions in the WS, i.e. the Session Execution, was not completely developed. A basic human interface to demonstrate the PU was also developed, to enable the execution of generation and detection sessions and to test the software. Through this interface the user can open and close the driver, test the memory, synchronise the time of both boards, download software, transfer data related with new sessions, control the execution of sessions and get results. Errors occurring during the generation and detection process are also reported by the basic interface.

6.7.4 Results

The approach used in the tests was to identify the actual generation and detection to obtain measurements. Parts of the final software were then downloaded into the boards developed and measurements carried out to validate the first ones.

In the generation task three main asynchronous and parallel processes can be identified: the "Calculate next packet", the "Transfer packet" and the "Transmit packet". The time consumed by the C25 to produce a packet is estimated in $t_{calc}+t_{pac}$ (Table 6.5) and the time to transmit a packet in $t_{pac}+t_{packet}$ (Table 6.6). The limits of the generation can be identified as the maximum packet processing rate of the C25, and the maximum packet transfer rate to the network.

Table 6.5 - Packet generation performance

Generated Function	t_{calc} us	t_{pac} us	Maximum Processing Rate $1/(t_{calc}+t_{pac})$ kpac/s
Two state	11	6	59
Markovian	7	6	77
Deterministic	6	6	83
AFAP	4	4	125

Table 6.6 - Packet transmission performance

Bandwidth (B) Mb/s	t_{packet} $(53\times8/B)$ us	t_{pac} us	Maximum Transfer Rate		
			PERC %	PERC×B Mb/s	PERC/t_{packet} kpac/s
10	42	6	87	9	21
20	21	6	78	15	37
30	14	6	70	21	50
60	7	6	54	32	77
150	2.8	6	32	48	114

$$PERC = t_{packet} / (t_{pac} + t_{packet})$$

Table 6.7 - Packet detection performance

Generated Function	t_{calc} us	$t_{arrival}$ us	Maximum Processing Rate $1/(t_{calc}+t_{arrival})$ kpac/s
No histograms	7	7	71
One histogram	17	7	42
Two histograms	28	7	29

The maximum processing rate has been calculated and measured for the general two state traffic profile and for other three representative traffic profiles. For all of them the packet length has been considered constant and equal to a cell (53 octets). The interpacket intervals for the Markovian profile are exponentially distributed and are constant for the deterministic function. For *AFAP*, packets are generated "*As Fast As Possible*". It should be mentioned that if packets with greater length are used, larger PERCs can be obtained and then larger packet transfer rates achieved. It can be said that the TGDS generation limit for a board is variable and depends on the **packet length**, the **generation function** and the **channel bandwidth**.

In the detection, as in the generation, three main asynchronous and parallel processes are identified: "Calculate", "Get arrival time" and "Receive packet". The limits of the detection can be identified as the maximum packet processing rate of the C25, and the maximum transfer rate from the network. The time consumed by the C25 to receive a packet is estimated by $t_{calc}+t_{arrival}$ (Table 6.7). The maximum processing rate has been calculated and measured for detection without histograms and with one and two histograms. Again, it can be said that the TGDS detection limit of a board is variable and depends on the **packet length**, the **measurements performed** and the **channel bandwidth**.

Analysing the results from generation and detection, the TGDS is limited by the Detector figures. However, since the two objectives of the TGDS were to validate the network trough the generation and detection of ATM cells, using this traffic over other traffic streams, and to generate variable load conditions of traffic well characterised for the demonstrator, we conclude that the results obtained are very good since they exceed all the initial requirements for the TGDS.

7 Experimental Results: Implementation of a Link between Transmitter and Receiver

A. Labrujere, O. Koning and P. Prinz

7.1 Demonstrator Set-up

During the evolution of the UCOL-project important results on both physical hardware and concepts for network management functions have been realized. Some of the hardware developed in the first two years of UCOL, has been used to realize a coherent transmission link. This demonstrator was built at PTT Research The Netherlands, and is composed of components developed by three partners. In this link the basic transmission scheme of the UCOL concept is demonstrated. The link, as indicated in Fig. 7.1, consists of:

- tunable narrow linewidth lasers followed by an active or passive phase modulator;
- a coherent DPSK receiver, consisting of a fibre based hybrid, low noise preamplifiers, and a demodulator;
- a comb generator for frequency stabilization;
- frequency locking of the frequency comb relative to a master laser.

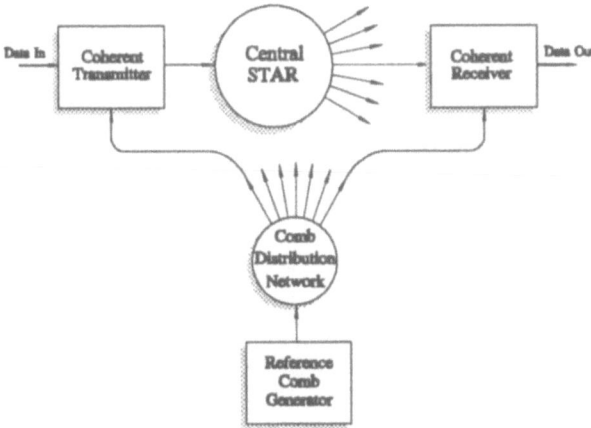

Fig. 7.1 - Coherent transmission link.

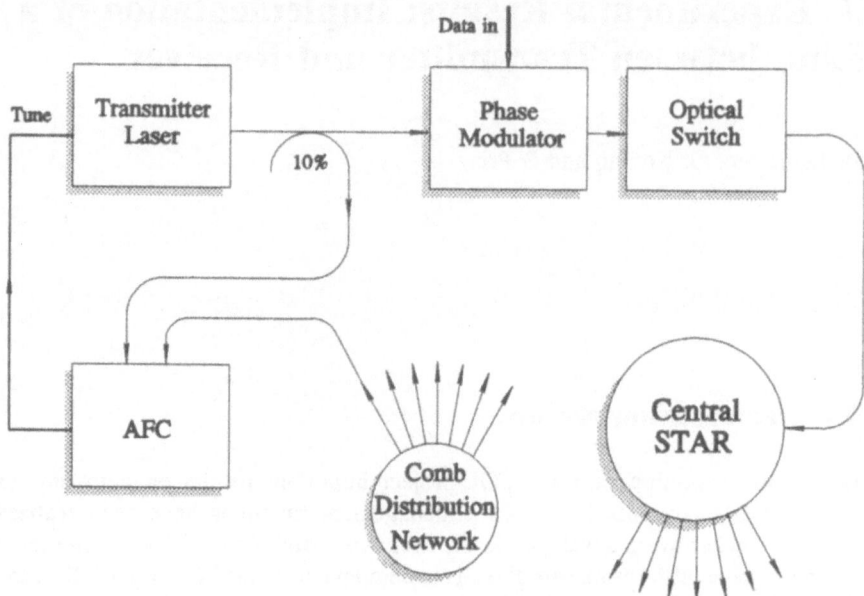

Fig. 7.2 - General set-up: coherent transmitter

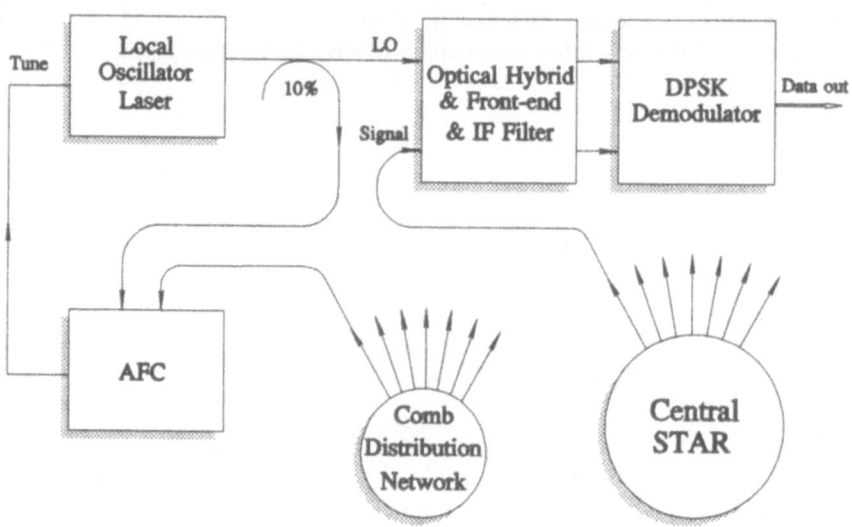

Fig. 7.3 - General set-up: coherent receiver ·

7.1.1 Transmission Link

Basically, the DPSK link consists of a coherent transmitter, the central star, and a coherent receiver. In the coherent transmitter (see Fig. 7.2) the data are modulated on an optical carrier. This optical carrier is generated by a tunable narrow linewidth laser. Since differential phase shift keying (DPSK) is chosen as modulation format, an external phase modulator is used. Optical switching, necessary to switch lasers off the network for TDMA, is not demonstrated in the present link. The central star is used to interconnect all transmitters and receivers in the network (maximally 1024). It is simulated by a variable attenuator. The coherent receiver (see Fig. 7.3) consists of a tunable narrow linewidth local oscillator laser and a receiver block. In the receiver block a balanced polarization diversity hybrid, frontends and a DPSK demodulator are integrated.

In the UCOL network, frequency stabilization by means of an optical reference frequency comb is applied. All transmitter and receiver lasers in the network can be locked in frequency relative to the various comb lines. For this purpose a small fraction of the transmitter or local oscillator laser power is tapped off and fed into a frequency control block (indicated as "AFC" in Figures 7.2 and 7.3).

For generation of the optical reference frequency comb a mode locked laser is used (see Fig. 7.4). In order to circumvent problems related to polarization mismatch of a reference line and the transmitter or L.O. laser, low speed polarization scrambling of the comb is applied.

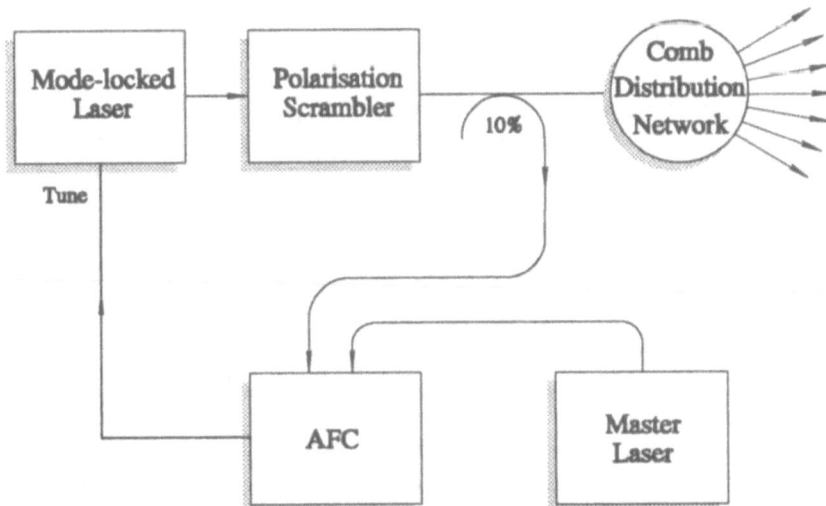

Fig. 7.4 - General set-up: reference comb generator block

To illustrate the possibility of locking the reference comb relative to an absolute frequency, the comb generator is locked to a master laser. By locking this master laser to an absolute frequency, absolute frequency control over the entire network is feasible.

7.1.2 Lasers

For the UCOL demonstrator tunable narrow linewidth lasers are needed. The requirement on linewidth for both transmitter and LO laser is 1 MHz or less. A continuous tuning range of several GHz is sufficient. In the demonstrator MQW-DBR lasers (*Multiple Quantum Well Distributed Bragg Reflector* lasers) supplied by OKI, are used. A 60 dB optical isolator is included in the laser module to avoid distortions of laser action by network reflections. The MQW-DBR laser consists of three sections: gain section, phase section, and Bragg section. Since a large increment of the linewidth (up to 15 MHz) is observed when current is provided to the Bragg and phase section, only the gain section is biased. For coarse (and slow) tuning the diode temperature can be varied. Fast tuning (over a limited range) can be applied by changing the current through the gain section. In the demonstrator a combination of current and temperature control is used.

7.1.3 Phase Modulators

During experiments two types of phase modulators were used:

Passive Phase Modulator. This was a pigtailed LiNbO$_3$ optical guided wave device, manufactured by Crystal Technology. The fibre-to-fibre insertion loss was 6.5 dB (note that at present LiNbO$_3$ devices with lower insertion losses are commercially available).

Active Phase Modulator. For this purpose a pigtailed semiconductor laser amplifier was used, which is manufactured by BT&D (type: SOA 3100). Both DC-current and RF-current were supplied to the amplifier by means of a bias-T.

Both modulators need a polarization controller at the input, since the performance is strongly polarization dependent. In order to shield the active phase modulator for external reflections, pigtailed isolators were used at both the input and the output fibre of the device.

7.1.4 Integration of Hybrid, Front-end and Amplifier

In the UCOL-concept, a DPSK polarization diversity receiver in a balanced configuration is used. The integrated receiver is schematically shown in Fig. 7.5 and consists of three modules:

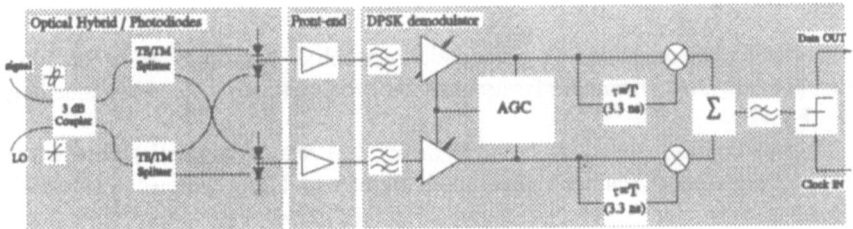

Fig. 7.5 - Electro-optics: optical hybrid, front-end and demodulator

- the polarization diversity hybrid and two balanced front-ends;
- an IF-filter;
- a DPSK demodulator.

Since polarization diversity is applied, the polarization of the signal light can be arbitrary. In order to have a contribution of the local oscillator power on both TE and TM paths of the receiver, it is necessary that the polarization of the local oscillator has both components. Therefore, a polarization controller is placed between the LO laser and the hybrid. In the hybrid, the light of both the signal and the local oscillator is multiplexed, and split in TE and TM components (see Fig. 7.5). Each branch is connected to a pair of balanced photodiodes. Behind the photodiodes, the electrical signals are amplified by low noise preamplifiers. After IF-filtering, the DPSK demodulation is performed in the last stage of the receiver.

For the front-end, two versions have been developed. The first untuned front-end is implemented in the transmission link. The two preamplifiers in the front-end have a noise current of respectively 5 and 6.5 pA/√Hz at 600 MHz. A better result is possible by using a tuned front-end: a value of 3.5 pA/√Hz has been obtained.

The IF-filters are mounted on a PCB and are characterized in combination with the front-ends. The measured transfer meets the UCOL specification on gain (>50 dB) and ripple (<1 dB). The centre of the filter transfer function is at about 605 MHz, only slightly above the specified 600 MHz.

7.1.5 Comb Generator

In UCOL a centrally distributed optical frequency comb is used for frequency stabilization of transmitter and receiver lasers in the network. A description of the comb generator and its operation can be found in Chapter 6, Section 6.1.

7.1.6 Polarization Scrambler

In the UCOL demonstrator a polarization scrambler has been used to facilitate frequency locking of tunable lasers in the network relative to the reference comb (see Fig. 7.4).

The polarization scrambler is based on *High Birefringent* (HiBi) fibre winded around a piezo tube. Standard single-mode fibres with FC/PC connectors are fusion spliced at both sides of the HiBi-fibre. A total optical power loss (including the FC/PC-connectors) of the device of 2 dB is measured.

If light is launched into the HiBi-fibre at a 45 degrees angle of polarization, the polarization at the output of the HiBi-fibre will change as a result of the radial expansion of the piezo tube. Since the comb polarization is scrambled at a rate faster than the AFC control speed, polarization diversity or control is not needed at the AFC receiver.

7.1.7 Automatic Frequency Control

In order to stabilize the relative frequencies of the lasers in the demonstrator, AFC-technology is applied. At three points in the link frequency control is necessary:

- to lock the comb to the master laser;
- to lock the transmitter laser at one side of a comb line;
- to lock the receiver laser at the other side of the same comb line.

It appeared that the AFCs work stably for a S/N-ratio >15 dB. Besides sensitivity, the frequency stability of the AFC is of great importance. To obtain maximal frequency stability, the frequency discriminator electrical input power needs to be constant. Therefore, some form of gain control is necessary. To avoid complex receivers, single detection instead of balanced detection has been used. Below the various AFCs in the transmission link are treated.

Transmitter and Receiver Laser Locking to Comb. A block diagram of the AFC used for locking of the transmitter and L.O. laser relative to a comb line is shown in Fig. 7.6. Both lasers need to be locked to a comb line in such a way that 600 MHz IF is achieved. AFCs with 300 MHz IF have been selected for both the LO laser and the transmitter laser. The 600 MHz difference between the LO laser and the transmitter laser can be achieved by relative locking to different sides of a comb line. In order to obtain a large S/N ratio, a high sensitive receiver is used and IF-filtering is applied. The IF-filter has a bandwidth of about 100 MHz; a compromise between a low noise bandwidth and a large locking range. A dynamic range of the IF-amplifier of about 20 dB has been targeted to ensure correct frequency discrimination for various input powers. For this reason, an AGC circuit providing this dynamic range has been included. For laser frequency control, both DBR-laser temperature and current could be adjusted. To enhance loop stability, the temperature control speed is set to several seconds. Fast frequency control (order msec) is provided by laser current control, supplied to the gain section.

Locking of the Comb to the Master Laser. A block diagram of the electronics used for frequency locking of the comb relative to the master laser is shown in Fig. 7.7. For comb frequency control, a comb line is locked in frequency to a strong master laser. The receiver sensitivity can be rather small, because the comb line power is at this point relatively high. For the automatic gain control, a limiting amplifier has been used. The frequency difference between a comb line and the master laser can be adjusted within a range of 200+500 MHz.

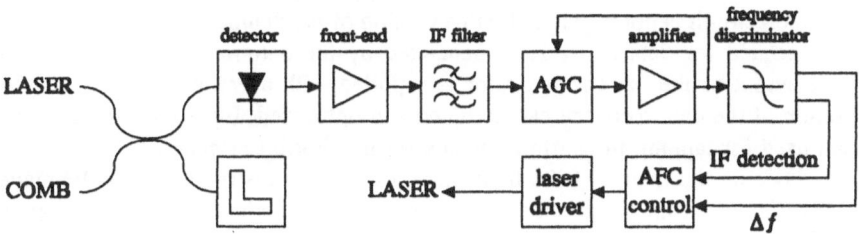

Fig. 7.6 - AFC: locking of laser to reference comb

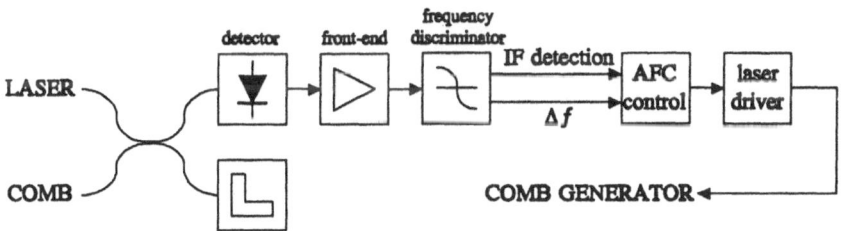

Fig. 7.7 - AFC: locking of comb to master laser

7.1.8 Optical Network

The UCOL network has been designed for the interconnection of maximally 1024 receivers and 1024 transmitters. The reference comb, needed for frequency stabilization of the network interfaces, is distributed to all receivers and transmitters on a different passive network.

In the demonstrator the distribution networks for signal and reference comb distribution are simulated by optical attenuators. A manual polarization controller is incorporated in the set-up to be able to measure polarization sensitivity. Since only very small optical power (-45+-65 dBm) is needed for the AFC controllers for stable locking the tunable lasers relative to the frequency comb, close attention must be paid to optical reflections in the network. Therefore, at some locations optical isolators are incorporated in the set-up.

7.1.9 BER Measurement Set-up

Adjustment of the decision level is provided by the DPSK demodulator. A clock extraction circuit is not provided by this version of the demodulator. Therefore, the clock needed for data recovery is generated by the pattern generator. Since the clock output of the pattern generator is supplied with a variable delay, synchronization of the data recovery can be tuned and optimized. Electrical splitters have been used for tapping the various signals for monitoring purposes.

In choosing the nature of the transmitted data a few restrictions must be made. Firstly, the spectrum of the modulated signal must contain small contributions at low frequencies.

Secondly, since a data generator is not included in the DPSK modulator, the transmitted and received bit sequence is not the same. Therefore, an alternative approach must be followed to circumvent this problem. The BER test-set has the possibility of transmitting and receiving user-selected bit sequences of adjustable bit lengths: therefore, by loading a sequence of bits at the data generator and the DPSK-transformed word at the error detector, BER measurements are still possible.

7.2 Experiments

In this Section, experimental results on the coherent transmission link are described. Firstly, the results on locking of lasers to the reference comb are given. Secondly, BER measurements on the transmission link are described for passive phase modulation. Some parameters affecting the transmission quality of the link are discussed. Finally, results on active phase modulation are presented.

7.2.1 Frequency Locking

In order to assure a constant IF in the AFC-loop, the power level at the input of the demodulator must be constant. Since automatic gain control is used in the AFC, a dynamic range of (optically) 20 dB can be accepted: the minimal comb line power for stable locking is -65 dBm, and the maximal power is -45 dBm (local

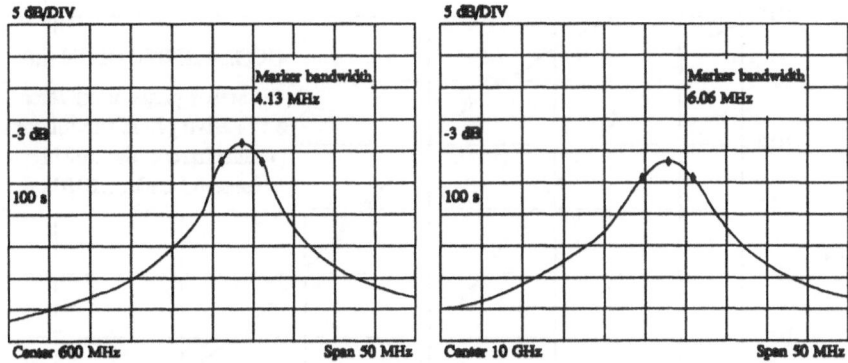

Fig. 7.8 - Left: beat signal of transmitter and LO lasers. Right: frequency difference between master laser and LO.

oscillator power available for frequency locking in this experiment was -12.3 dBm).

To obtain proper demodulation, a stable intermediate frequency is of vital importance; therefore, the stability of the intermediate frequency has been measured. In the locking set-up, both lasers were locked relative to a comb line. Both IFs to the comb line were 300 MHz, and since both lasers were allocated at different sides of the comb line, the total IF is about 600 MHz. In Fig. 7.8 the beat signal of the transmitter and local oscillator laser, as obtained on an electrical spectrum analyzer is shown. One can see in this figure that the long term stability (100 sec average time) of the intermediate frequency is about 4.2 MHz (FWHM).

In a second experiment the stability of locking the comb relative to the master laser has been measured. Therefore, the frequency difference between the master laser and the local oscillator (locked to the comb) has been determined. The comb is locked to the master laser and the L.O. laser is locked to a comb line. Fig. 7.8 shows a frequency difference between both sources of about 10 GHz, because the master laser and the L.O. laser are locked to different comb lines (spacing comb lines is about 4.72 GHz). The 3 dB bandwidth (FWHM) of the frequency difference, averaged over 100 sec, is 6.1 MHz.

7.2.2 DPSK Transmission Link

In a first set of experiments the performance of the DPSK link is tested using both the LiNbO$_3$ passive phase modulator and the semiconductor active phase modulator. The BER as function of the receiver input power is measured. The optical power budget in case of active and passive phase modulation has been determined. Further, the dependence of the BER as a function of the intermediate frequency, modulation depth, clock delay and receiver input SOP are determined.

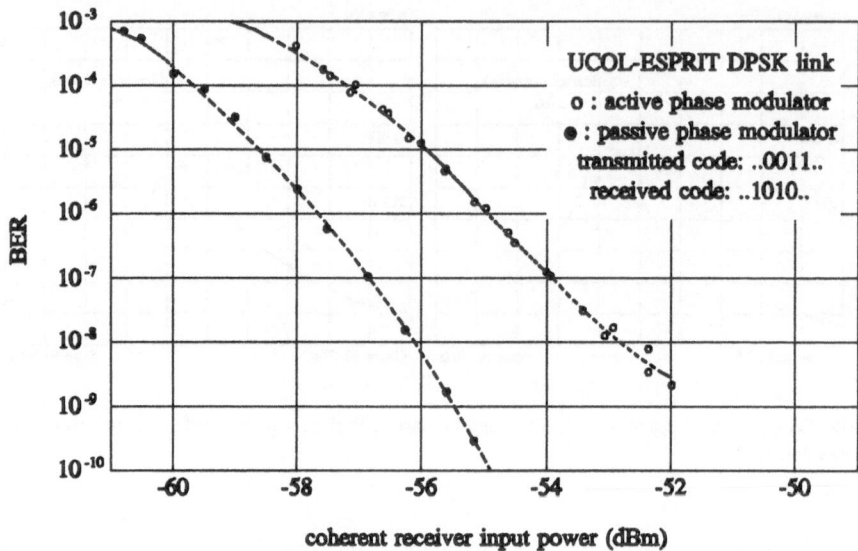

Fig. 7.9 - DPSK link: measured BER using either a passive or an active phase modulator

Note the transmitted bit sequence, which is shown in Fig. 7.9.

Passive Phase Modulation. The BER has been measured as function of the optical signal power, incident at the coherent receiver (see Fig. 7.9). From this measurement a receiver sensitivity of -57.5 dBm at a BER of 10^{-6} has been found. The maximal optical power at the output of the phase modulator was measured to be -9.3 dBm.

Active phase modulation. An active phase modulator is implemented in the DPSK link. A measurement of BER versus coherent receiver input power is also shown in Fig. 7.9. The polarization at the input of the BT&D semiconductor laser amplifier has been adjusted to provide maximal output power. The amplifier was operating at 20°C, and at a DC-current of 100 mA. An optical output power of the amplifier (including an optical isolator) of +4.8 dBm was measured. From the measurement a sensitivity of -54.65 dBm (at BER = 10^{-6}) has been found. When compared to passive phase modulation a degradation of 2.9 dB has been found.
It must be noted that the laser amplifier's temperature has not been optimized for optimal phase modulation. From earlier measurements it is found that the ratio of *Phase Modulation* (PM) versus *Amplitude Modulation* (AM) is strongly dependent on the position of the gain peak relative to the amplifier gain ripple [4]. This may be used to decrease the penalty due to spurious intensity modulation.

Power Budget. In the case of active phase modulation a degraded receiver sensitivity has been measured. On the other hand, the transmitter output power is

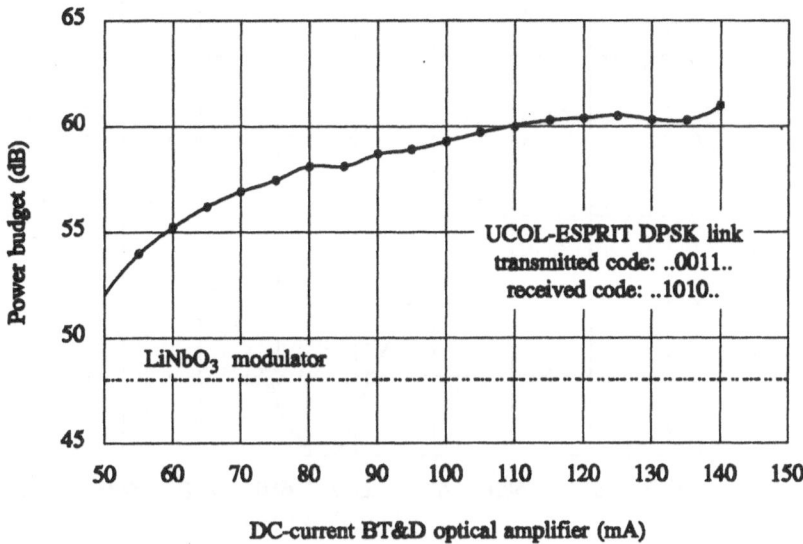

Fig. 7.10 - UCOL power budget with the active phase modulator

much larger in the active when compared to the passive approach. This is illustrated in Fig. 7.10 where the optical power budget is shown as a function of the active phase modulator dc-current. For comparison, the power buget in the passive approach is also shown. One can see that an increase of 12 dB in power budget is feasible when using active phase modulation.

Influence IF. In Fig. 7.11 the penalty as function of the intermediate frequency (IF) is shown. A penalty of less then 1 dB is achieved for an IF mismatch of ±7 MHz. Since the error detection level of the AFC-frequency discriminator can be changed manually, the IF can be adjusted. It is shown that the optimal IF is slightly higher than 600 MHz, reflecting the properties of the IF filter.

Modulation depth. The influence of the modulation depth has also been determined. In Fig. 7.12 the modulator's drive voltage (in arbitrary units) is shown. Again the penalty on receiver sensitivity has been determined.

Clock timing. In Fig. 7.13 the penalty of receiver sensitivity on clock-delay is shown. From this measurement the required accuracy of the clock extraction circuit can be determined to be ±0.25 ns for a penalty < 0.5 dB.

Local oscillator power. The receiver sensitivity has been measured as a function of local oscillator laser power. The measurements are shown in Fig. 7.14: the SOP of both local oscillator and signal input and were adjusted to either TE or TM

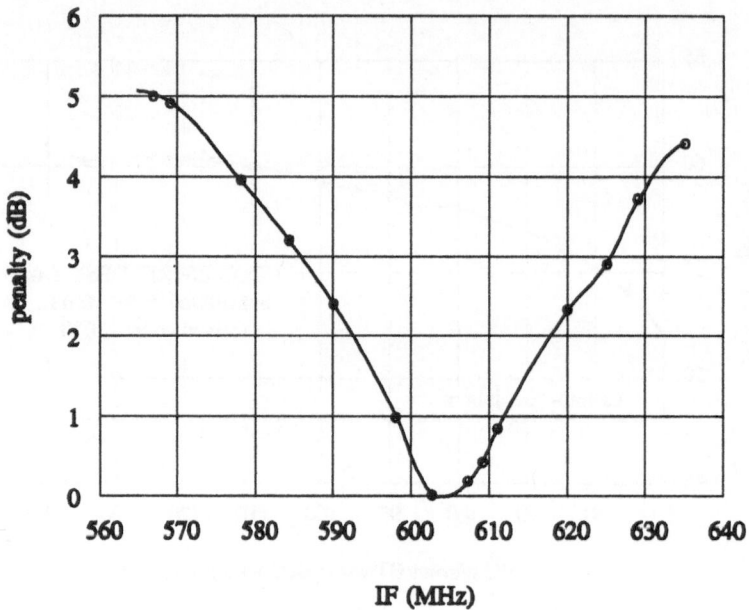

Fig. 7.11 - Influence of an IF-shift (LiNbO$_3$ phase modulator)

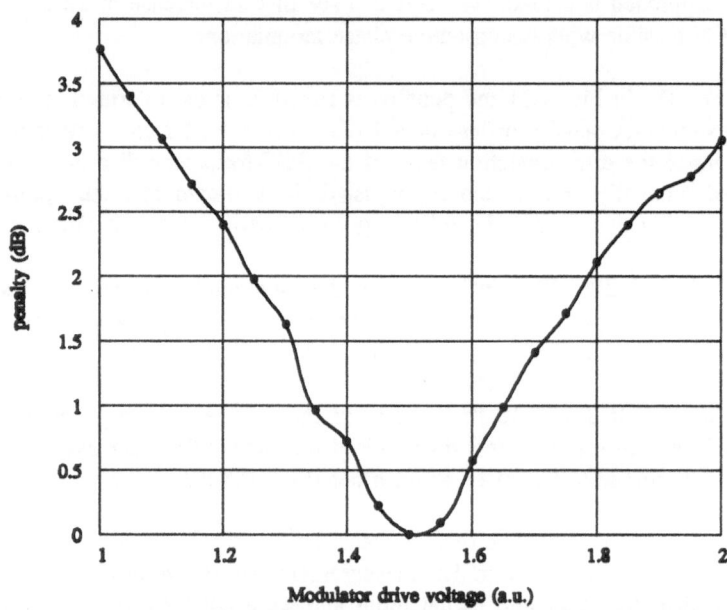

Fig. 7.12 - Influence of the modulation depth (LiNbO$_3$ phase modulator)

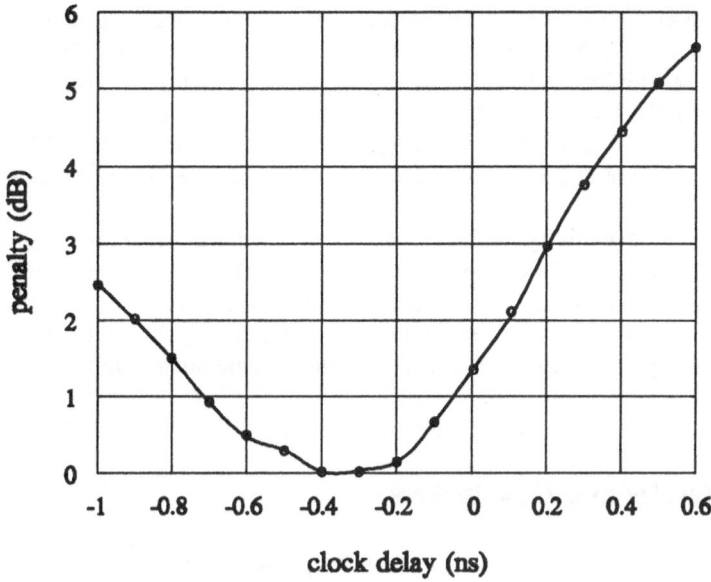

Fig. 7.13 - Influence of the clock delay (LiNbO$_3$ phase modulator)

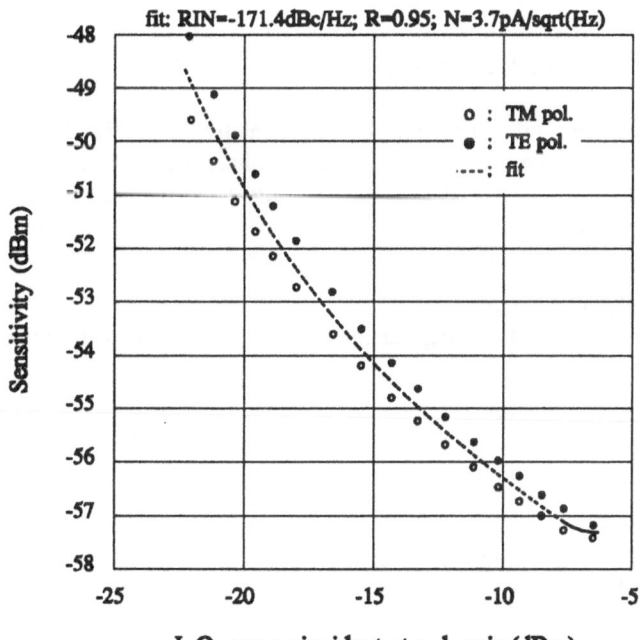

Fig. 7.14 - Influence of LO power

polarization. The difference in sensitivity for the two branches is dependend on local oscillator power, and about 1 dB. The influence of LO laser power on receiver sensitivity can be calculated, taking into account the receiver noise and the LO laser RIN. From a fit (also shown in the figure) the average receiver noise is estimated at 3.7 pA/√Hz, and the effective RIN is found to be -171.4 dBc/Hz. These values are in good agreement with direct characterization measurements.

Polarization sensitivity. A 0.7 dB variance in receiver sensitivity, depending on the SOP of the signal input, has been measured. This can be ascribed to different optical losses in TE and TM branches of the hybrid (0.13 dB), front-end noise current, asymmetric LO distribution, or a variability in performance of the electrical signal branches (front-end, IF filters, demodulator etc.)

7.3 Results and Conclusions

In this Section the receiver sensitivity of the present coherent receiver is calculated and compared with values from system calculations. It must be noted that only a few experiments have been performed up to now. Improvement on receiver sensitivity might be possible in the future.

7.3.1 Passive Phase Modulation

In this Section an estimation of the sensitivity of the DPSK-receiver is calculated. Firstly, the case where a lithium-niobate phase modulator is used in the coherent network interface is detailed. The ideal receiver sensitivity for a DPSK polarization diversity receiver at a BER of 10^{-6} is -61.5 dBm. For the non-ideal case a number of penalties on the receiver sensitivity are expected:

- Equivalent noise bandwidth:
  For N×BW = 1.5×BR a degradation of the receiver sensitivity of 1.8 dB is estimated.

- Photodiode responsivity:
  The responsivity of the photodiodes in the present set-up is 1 A/W. This results in a 0.9 dB penalty.

- Phase noise:
  The penalty originating from the laser phase noise ($\Delta v = 1$ MHz) is estimated at < 1.1 dB

- IF shift:
  An IF shift of ±3 MHz corresponds to a 0.3 dB penalty in receiver sensitivity.

- Hybrid excess loss:
  As the excess loss of the fibre-optic hybrid is estimated to be 0.3 dB, it results in a 0.3 dB penalty.

- TE/TM extinction ratio:
  For the fibre optic hybrid a -20 dB extinction ratio has been measured. This results in a 0.2 dB penalty.

- Thermal noise:
  For the preamplifier used during experiments a noise current is specified at 3.5 pA/√Hz. The penalty on receiver sensitivity is estimated at 0.8 dB.

- Laser RIN:
  From measurements a laser RIN of -150 dBc/Hz is found. The effective RIN, experienced at the balanced receiver input is -170 dBc/Hz. The penalty on receiver sensitivity based on this value is far below 0.1 dB.

7.3.2 Active Phase Modulation

By using active instead of passive phase modulation, a receiver sensitivity of -54.65 dBm is found. This is 2.9 dB worse than the situation in which passive phase modulation was used. This difference in sensitivity can be explained by the effect of spurious intensity modulation. If the penalty is completely due to intensity modulation, we can calculate the intensity modulation index to be 0.74, and the linewidth enhancement factor β^* to be 4.2. This rather low value has been observed as a result of the ripple of modulation response [4]. Note that the penalty is rather high compared to publications in literature. A better result is expected, when fine tuning of the laser frequency with respect to the ripple of the modulation response is apllied. Moreover, by using multi-section optical amplifiers, the penalty can be decreased up to about zero.

In the DPSK demonstrator a simple bit sequence was applied, and the set-up consists of only one transmitter and one receiver. In a realistic system however a number of extra penalties will occur:

- *Non-flat PM response:* if transmitting a random bit sequence the modulator's phase modulation response needs to be flat. By using a equalizing circuit at the phase modulator input the penalty can be reduced to zero.
- *Channel crosstalk:* at a channel spacing of 1200 MHz the penalty due to channel crosstalk is estimated at 0.5 dB.
- *Clock jitter:* at the experiments the BER test-set was used as a clock generator source. If a clock recovery circuit is used, the sensitivity degradation induced by the clock jitter is estimated at < 0.25 dB for a standard deviation in jitter of < 0.3 ns.
- *Spontanuous emission active phase modulators:* for the present network concept at the same time maximally 20 channels are transmitting signals into the star. The coherent receiver spontaneous emission generated by 20 channels will degrade the receiver sensitivity. The penalty is estimated to be 0.7 dB.

- *Low-frequency content of data:* the phase modulator's response at low frequencies (several MHz) indicates that transmission of data having large low-frequency content seems difficult. Therefore some kind of coding may be needed to minimize the penalty.

7.3.3 Comparison between Active and Passive Phase Modulation

Although the receiver sensitivity for passive modulation is 2.9 dB higher than that for active phase modulation, the power budget for the active approach is larger. For passive phase modulation the budget is 48 dB, whereas for the active approach a power budget of over 60 dB has been found. This difference of 12 dB is caused by optical amplification in the semiconductor laser amplifier.

It must be noted that the passive modulator shows relatively high insertion losses (6.5 dB). Presently, devices having 3 dB attenuation are available. However, since an on/off switch of 50 dB isolation must be placed behind the modulator, an extra insertion loss of minimal 3.5 dB must be expected. Therefore, the above calculated power budget increase seems realistic.

7.3.4 Conclusions

At PTT Research, a coherent transmission link, consisting of components developed in the first two years of the UCOL project, has been realized. In the system, the feasibility of:

- stabilization to a reference frequency comb generated by a mode-locked laser,
- active phase modulation,
- and reception by a DPSK receiver,

has been tested. The results of the transmission link are promising. Firstly, long term stabilization with respect to the comb lines has been demonstrated. Secondly, the DPSK receiver has been tested by using a LiNbO$_3$ phase modulator. This receiver consists of a fiber based hybrid, two low noise preamplifiers, and a DPSK-demodulator. The receiver sensitivity has been determined to -57.5 dBm.

The third aspect concerns active phase modulation, of which the feasibility has also been demonstrated. Active phase modulation introduces a penalty of 2.9 dB compared to passive modulation. This penalty must be ascribed to spurious intensity modulation, and possible to non-perfect modulation response. The optical amplification process provides however a possible system expansion increase of 12 dB. This improved power budget can be increased even further, if better amplifiers are used. Moreover, at present it has been only possible to do a few link characterization experiments due to a shortage of time. It is expected that by tuning the laser wavelength relative to the gain ripple, the penalty of intensity modulation can be decreased. Moreover, the intensity modulation penalty can almost completely be avoided, if two-section amplifiers are being used. In such a system, an increase of more than 15 dB in power budget is feasible.

8 UCOL Bibliography

8.1 Technical Reports

Report on external cavity module. (GEC, 5-Apr-1990)

Report on tunable narrow-linewidth laser. (GEC, 18-Oct-1990)

Report on optimization of diode pumped Er^{3+}-doped fibre laser. (UoS, 9-Feb-1990)

Report on narrow-linewidth operation of Er^{3+}-doped fibre lasers. (UoS, 6-Jun-1990)

Er^{3+}-doped fibre optimisation, EDFA phase and amplitude noise measurements and performance. (Uos, 30-Nov-1991)

Interim report on Act. 1.3.1: "System study of multi-line master laser". (PTT-RNL, 30-Aug-1989)

Final report on Act. 1.3.1: "System study of multi-line master laser". (PTT-RNL, 1-Mar-1990)

Report on Act. 1.3.3: "Realisation of the multi-line source". (PTT-RNL, 1-Mar-991)

Report on analysis of operational performance. (PTT-RNL, 1-Dec-1991)

Report on characterization of first samples of optical couplers and availability of two samples. (GEC, 29-Nov-1989)

Report on design of loop electronics. (GEC, 6-Oct-1989)

Report on Act. 2.2.1: "Automatic frequency control circuitry". (GEC, 18-Oct-1990)

Report on design of frequency offset electronics. (GEC, 30-Aug-1989)

Report on Act. 2.3.2: "Frequency offset: construct circuit" (GEC, 23-Oct-1990)

Report on Act. 2.4.1: "Controller design". (GEC, 23-Oct-1990)

Report on assessment of high isolation switch structures. (GEC, 9-Feb-1990)

Design of an electrode test mask for efficient optical comb-line generation. (GEC, 10-Jan-1991)

Report on Act. 3.4.1: "Passive phase modulator". (FACE, 31-May-1991)

Interim report on Act. 3.4.2: "Active phase modulator and switch". (PTT-RNL, 1-Sep-1990)

Final report on active phase modulator and switch. (PTT-RNL, 1-Mar-1991)

Report on Act. 3.4.3: "Active phase modulator characterisation (in transmission link)". (PTT-RNL, 1-Dec-1991)

Progress report on the study of the 3 dB coupler. (AA-RC, 5-Mar-1990)

Progress report on the study of TE/TM splitter. (AA-RC, 5-Mar-1990)

Report on complete characterization of the first sample of photodiode array. (AA-RC, 5-Mar-1990)

Report on Act. 4.1.5: "Technology improvement (of the $LiNbO_3$ polarisation diversity hybrid)". (AA-RC, 4-Apr-1991)

Report on development of the polarization diversity hybrid. (AA-RC, 4-Apr-1991)

Report on wavelength window optimization of fibreoptic TE/TM splitters. (DBR, 4-Dec-1989)

Report on reproducibility of peak wavelength. (DBR, 14-Mar-1990)

Report on splicing techniques for polarization maintaining fibers. (DBR, 29-Nov-1989)

Report on housing technique and integration of TE/TM splitters. (DBR, 7-Jan-1991)

Report on epitaxial layer growth for PIN-photodiode pair. (DBR, 28-Feb-1990)

Report on structuring techniques. (DBR, 28-May-1990)

Report on mounting and connection techniques. (DBR, 13-Mar-1991)

Report on Act. 4.3.1: "Preamplifier development (first prototype)". (PTT-RNL, 14-Sep-1990)

Report on Act. 4.3.2 - Prototype of receiver preamplifier without tuning. (PTT-RNL, 12-Apr-1991)

Report on demodulator electronics and test in continuous mode. (FACE, 31-May-1991)

Report on Act. 4.3.6: "Design of a TDMA test signal generator". (FACE, 31-May-1991)

Report on the modeling activity. (FACE, 14-Sep-1990)

Report on Act. 4.3.10 - Measurements on AA-RC and DBR photodiodes. (PTT-RNL, 30-Aug-91)

Report on tuning. (PTT-RNL, 1-Dec-1991)

Fibre pigtailed GRIN-lens Fabry-Perot type filter. (SEL, 27-Aug-1990)

Prototype of tunable optical filter. (SEL, 28-Feb-1991)

All-fibre polarization control for manual adjustment (SEL, 28-Aug-1989)

Derivation of error signals for automatic SOP control. (SEL, 23-Feb-1990)

Network expansion by optical amplifier. (SEL, 8-Mar-1991)

EDFA analysis and specifications. (SEL, 28-Mar-1991)

Polarization scrambler characterization. (SEL, 28-Feb-1991)

EDFA pump source and hardware configuration test. (SEL, 13-Dec-1991)

Report on system assessment. (FACE, 31-Aug-1989)

Development tools choice and study. (TESA, 8-Aug-1990)

Report on DQAC protocol specifications. (FACE, 19-Jun-1990)

Report on system start-up and restart strategy (FACE, 27-Sep-1990)

Report on Objects definition. (COSI, 21-Sep-1990)

Report on distance measurement strategy. (FACE, 21-Sep-1990)

Considerations on Management workstation specifications. (INESC, 15-Mar-1991)

First draft on MMI specifications. (INESC, 19-Mar-1991)

Report on the G.703 34 Mb/s interface HW specifications. (INESC, 31-Mar-1991)

Report on DQAC error correction logic design in the Information field. (FACE, 28-Sep-1990)

Report on DQAC error correction logic design in the Queue Status field. (FACE, 27-Sep-1990)

Report on Video Interface. (INESC, 26-Apr-1990)

Report on the 34 Mb/s TA specifications. (INESC, 31-Mar-1991)

Report on AAL/Synchronisation test system. (INESC, 30-Nov-1991)

Report on Synchronisation Mechanisms testing strategies. (INESC, 30-Nov-1991)

Report on the activity with specifications of Generator/Detector station and interfaces for station and workstation. (INESC, 31-Mar-1990)

Interim report on specifications and development of TGD and measurement station. (INESC, 31-Mar-1991)

Interim report on Act. 10.2.3: "Traffic generator/detector station SW development". (INESC, 31-Mar-1991)

Processing unit (hardware). (INESC, 29-Nov-1991)

Processing unit (software) and report on experiments. (INESC, 29-Nov-1991)

Interim report on Market approach. (IDATE, 9-Nov-1989)

Report on documentary study. (IDATE, 9-May-1990)

8.2 Publications

W.L. Barnes, D.J. Taylor, M.E. Fermann, J.E. Townsend, L. Reekie and D.N. Payne: "A Diode-Laser-Pumped, Er^{3+}-Doped Fibre Laser Operating at 1.57 um" - Proc. OFC'89, Houston, USA, Technical Digest, Paper TUG4, February 1989

S.B. Poole and L. Reekie: "Tunable Single-Mode Fibre Lasers" - Proc. ILA, Tokyo, Japan, January 1989

W.L. Barnes, L. Reekie and D.N. Payne: "Highly Efficient Er^{3+}-Doped Fibre Lasers Pumped at 980 nm" - Proc. IOOC'89, Kobe, Japan, Paper 20A3-3, July 1989

P.R. Morkel and R.I. Laming: "Theoretical Modeling of EDFAs with Excited State Absorption" - Optics Letters, Vol. 14, N. 19, pp. 1062-1064, 1989

R.I. Leming, L. Reekie, P.R. Morkel and D.N. Payne: "Multichannel Crosstalk and Pump Noise in Characterisation of EDFA Pumped at 980 nm" - Electronics Letters, Vol. 25, N. 7, pp 455-456, 1989

W.L. Barnes, P.R. Morkel, L. Reekie and D.N. Payne: "High Quantum Efficiency Er^{3+}-Fibre Lasers Pumped at 980 nm" - Optics Letters, Vol. 14, N. 18, pp 1002-1004, 1989

W.L. Barnes, S.B. Poole, J.E. Townsend, L. Reekie, D.J. Taylor and D.N. Payne: "Er^{3+}-Yb^{3+}- and Er^{3+}-Doped Fibre Lasers" - Journal of Lightwave Technology, Invited Paper, (Special Issue OFC'89), Vol. 7, N. 10, pp. 1461-1465, 1989

D.N. Payne, R.I. Laming, W.L. Barnes, L. Reekie and P.R. Morkel: "Erbium-Doped Fibre Lasers and Amplifiers" - Proc. SPIE Conference, "Fibre Laser Sources and Amplifiers", Vol. 1171, 1989

D.P. Hand and P.St.J. Russell: "Photoinduced Refractive Index Changes in Germanosilicate Fibres" - Optics Letters, Vol. 15, N. 2, pp 102-104, 1990

D.P. Hand, P.St.J. Russell and P.J. Wells: "UV-Induced Refractive Index Changes in Germanosilicate Fibres" - Proc. "Photorefractive Materials Meeting", OSA Topical Meeting "Photorefractive Materials, Effects and Devices", Aussois, France, 1990

D.P. Hand, L.J. Poyntz-Wright and P.St.J. Russell: "Enhanced Photorefractivity in Germanosilicate Fibres: Effects of Bleaching with 488 nm Light" - Proc. Integrated Photonic Research, Vol. 5, Paper MJ3, Hilton Head, South Carolina, USA, 1990

G.J. Cowle, P.R. Morkel, R.I. Laming and D.N. Payne: "Spectral Broadening Due to Fibre Amplifier Phase Noise" - Electronic Letters, Vol. 26, N. 7, pp 424-425, 1990

P.R. Morkel, G.J. Cowle and D.N. Payne: "Single-Frequency Operation of a Traveling-Wave Erbium Fibre Ring Laser" - Proc. OFC'90, San Francisco, USA, 1990

D.N. Payne, R.I. Laming and G.J. Cowle: "Performance of Erbium-Doped Fibre Amplifiers" - Proc. 35th Micro-Optics Meeting, Invited Paper, Tokyo, Japan, 1990

P.R. Morkel, G.J. Cowle and D.N. Payne: "A Traveling-Wave Erbium Fibre Ring Laser with 60 kHz Linewidth" - Electronic Letters, Vol. 26, N. 10, pp 632-634, 1990

R.I. Laming, D.N. Payne and G.J. Cowle: "Applications of Erbium-Doped Fibre Amplifiers in Communications" - Proc. of 8th International Conference on Fibre Optics and Optoelectronics, London, UK, 1990

G.J. Cowle, P.R. Morkel, R.I. Laming and D.N. Payne: "Spectral Broadening due to Fibre Amplifier Phase Noise" - Proc. of 2nd Bangor Communications Symposium, University of Wales, UK, 1990

M. Tachibana, R.I. Laming, P.R. Morkel and D.N. Payne: "Gain-Shaped Erbium-Doped Fibre Amplifier with Broad Spectral Bandwidth" - Proc. of Topical Meeting on Optical Fibre Amplifiers and their Applications, Paper MD1, Monterey, California, USA, 1990

G.J. Cowle, L. Reekie, P.R. Morkel and D.N. Payne: "Narrow Linewidth Fibre Laser Sources" - Proc. of SPIE OE/Fibres Conference, Invited Paper, San Jose, California, USA, 1990

M. Tachibana, R.I. Laming, P.R. Morkel and D.N. Payne: "Erbium-Doped Fibre Amplifier with Flattened Gain Spectrum" - Photonic Technology Letters, 1990

G.J. Cowle, D.N. Payne and D. Reid: "Single-Frequency Traveling-Wave Erbium-Doped Fibre Loop Laser" - Electronic Letters, Vol. 27, N. 3, pp 229-230, 1991

G.J. Cowle, D.N. Payne and D. Reid: "Traveling-Wave Erbium-Doped Fibre Loop Laser" - IEE Colloquium "Sources for Coherent Communications", Paper 9, London, UK, 1991

A. Fioretti, C.A. Rocchini, L. Torchin and S.R. Treves: "Fibre Optic LAN/MAN Architectures for Coherent Optical Transmission" - Electrical Communications, Vol. 62, N. 3/4, 1988

S.R. Treves: "Broadband Local and Metropolitan Area Networks in an IBCN Environment" - Proc. of ICCC '88, Tel Aviv, Israel, 1988

A. Fioretti, E. Neri, S. Forcesi, A.E. Green, P.N. Fernando, A.C. Labrujere, O.J. Koning, J.P. Bekooij, G. Veith and H. Schmuck: "Technology Aspects of a Coherent Optical MAN." - Proc. of SPIE, Vol. 1175, Boston, Massachusetts, USA, 1989

A. Fioretti and S. Forcesi: "UCOL: a Concept for an Ultra-Wideband Coherent Optical MAN" - Proc. of OCTIMA International Workshop, Roma, Italy, 1989

D. Capolupo, A. Fioretti, S. Forcesi and E. Neri: "UCOL: Evolution of the System Concept During the Realisation Phase" - Proc. of 6th Annual ESPRIT Conference, Brussels, Belgium, 1989

A. Fioretti, E. Neri, S. Forcesi, A.C. Labrujere, O.J. Koning and J.P. Bekooij: "Research on Coherent Optical LANs" - Proc. of 4th Tirrenia International Workshop on Digital Communications, Tirrenia, Italy, 1989

A. Fioretti, E. Neri, S. Forcesi, O.J. Koning, A.C. Labrujere, J.P. Bekooij, B. Hillerich, E. Weidel and G. Veith: "An Evolutionary Configuration for an Optical Coherent Multichannel Network" - Proc. of GLOBECOM'90, San Diego, California, USA, 1990

D. Capolupo and F. Del Castello: "A Flexible Integrated Multiservice Optical Coherent Network based on TDMA Techniques" - Proc. of EFOC-LAN'90, Munich, Germany, 1990

A. Fioretti and S.R. Treves: "A Novel Distributed Photonic Switch" - Proc. of ISS'90, Stockholm, Sweden, 1990

S.R. Treves: "Trends and Challenges in Photonic Switching" - Proc. of ITS'90, Taipey, Rep. of China, 1990

A. Bianchi, G. Tolusso and S.R. Treves: "Automatic Reconfiguration of Multichannel Networks" - Proc. of TELECOM 1991 Technical Symposium, Geneve, Switzerland, 1991

R. Deufel, M. Eisenmann, H. Gottsmaann, B. Hillerich, A. Schurr and E. Weidel: "All-Fiber Optical Front-end for a Balanced Polarisation Diversity Receiver with 0.2 dB Excess Loss and High Noise Rejection" - Submitted for publication to Journal of Optical Communications, 1991

O.J. Koning, A.C. Labrujere, C.M. de Block and J.P. Bekooij: "Reference Frequency-Comb for Multi-Channel Stabilisation by Mode-Locking of a Semiconductor Laser" - Proc. of ECOC'90, Amsterdam, NL, 1990

J.P. Bekooij, O.J. Koning and A.C. Labrujere: "Carrier Stabilisation in Coherent Multi-Channel Systems using a Comb of Reference Frequencies" - Proc. of OCTIMA'91, Roma, I, 1991

A.C. Labrujere, C.A.M. Steenbergen and C.J. van der Laan: "Phase Modulation and Optical Switching by a Semiconductor Laser Amplifier" - Proc. of 2nd Topical Meeting on Optical Amplifiers and their Applications, Colorado, USA, 1991

H. Schmuck and G. Veith: "Low-Loss, SM Fibre-pigtailed GRIN-Lens Fabry-Perot Type Filter for Widely Variable Free Spectral Range" - Proc. of ECOC'90, Amsterdam, NL, 1990

G. Veith, H. Schmuck and Th. Pfeiffer: "Advanced Fibre Based Components for OFDM Systems" - Proc. of OCTIMA'91, Rome, I, 1991

Th. Pfeiffer and H. Schmuck : "Widely Tunable Actively Mode-Locked Erbium Fibre Ring Laser" - Proc. of 2nd Topical Meeting on Optical Amplifiers and their Applications, Snowmass Village, Colorado, USA, 1991

H. Schmuck and Th. Pfeiffer: "Fibre Pigtailed Fabry-Perot Filter Used as Tuning Element and for Comb Generation in an Erbium-Doped Fibre Ring Laser" - Proc. of ECOC'91, Paris, France, 1991

H. Schmuck, Th. Pfeiffer and G. Veith : "Widey Tunable Narrow Linewidth Erbium Doped Fibre Ring Laser" - El. Letters, Vol. 27, n. 23, pp 2117-2118, 1991

J.R. Wilcox: "OFDM Comb Generation using Synchronous Traveling Wave Phase Modulators" - Proc. of SPIE Conference, "Advanced Optoelectronic Technology, Vol. 864, 1987

A.D. Carr: "A High Isolation Switch in $LiNbO_3$ for Coherent Optical LANs" - IEE Colloquium "Coherent Optical Communications", Digest N. 1990/116, London, UK, 1990

P.N. Fernando, M. Fake and A.J. Seeds: "A Novel Approach to Optical Frequency Synthesis in Coherent Lightwave Systems" - Proc. of SPIE Conference, "Coherent Lightwave Communications", Vol. 1372, 1990

P.N. Fernando, R.T. Ramos and A.J. Seeds: "Optical Carrier Synthesis in Coherent Ligthwave Systems" - IEE Colloquium "Sources for Coherent Optical Communications", London, UK, 1991

C. Duchet and N. Flaaronning: "New TE/TM Polarisation Splitter made in Ti:$LiNbO_3$ using X-cut and Z-axis Propagation" - Electronic Letters, Vol. 26, N. 14, pp 995-996, 1990

C. Duchet, N. Flaaronning, C. Brot and L. Sarrabay: "A New Integrated Optic TE/TM Splitter made on $LiNbO_3$ Isotropic Cut" - Proc. of SPIE Conference "Coherent Lightwave Communications", Vol. 1372, 1990

E. Carrapatoso, C. Aguiar and M. Ricardo: "Traffic Generator/Detector System Concept" - Proc. of EFOC LAN '90, Munich, Germany, 1990

M. Ricardo, E, Carrapatoso and C. Aguiar: "A Traffic Generator/Detector System for High Speed Networks" - Bilkent International Conference on New Trends in Communication, Control and Signal Processing, Bilkent University, Ankara, Turkey, 1990

C. Aguiar, E. Carrapatoso, M. Ricardo, F. D'Ignazio and A. Fantini: "The Usage of Prediction in Demanded Assignment with Distributed Control Protocols" - Bilkent International Conference on New Trends in Communication, Control and Signal Processing, Bilkent University, Ankara, Turkey, 1990

J.N. Almeida, J.M. Cabral, M. Ricardo, E. Carrapatoso and A.P. Alves: "Integrated Multi-Services Network Using ATM and Coherent Optical Techniques" - Proc. of ENDIEL'91, Lisbon, Portugal, 1991

J.N. Almeida, J.M. Cabral and A.P. Alves: "End-to-End Synchronisation in Packet Switched Networks" - Proc. of ACM - 2nd International Workshop on Network and Operating System Support for Digital Audio and Video, Heidelberg, Germany, 1991

References

1. Rosenberg: Data Communications - September '88, pp.71–78
2. Hindus: Lightwave - May '89, p. 16A.
3. A. Saleh et al., IEEE Photon Technol. Lett. **2** (1990) pp. 714-717
4. Labrujere, A.C., Steenbergen, C.A.M., van der Laan, C.J.: *Phase Modulation and Optical Switching by a Semiconductor Laser Amplifier*. Proc. of 2nd Topical Meeting on Optical Amplifiers and Their Applications, Snowmass Village, Colorado, 1991, paper THC4

UCOL Consortium

PARTNER	ACRONYM	ADDRESS
Alcatel FACE S.p.A Research Centre	FACE	Viale L. Bodio 33 20100 Milano, I
SEL Alcatel Research Centre	SEL	Holderaeckerstrasse 35 D-7000 Stuttgart 31, D
Koninklijke PTT Nederland PTT Research	PTT-RNL	Sint Paulusstraat 4 2264 XZ Leidschendam, NL
University of Southampton	UoS	University of Southampton Southampton SO9 5NH, UK
Daimler-Benz Research Centre	DBR	Wilhelm-Runge-Strasse 11 D-7900 Ulm, D
Alcatel Alsthom Recherche	AA-RC	Route de Nozay 91460 Marcoussis, F
GEC Hirst Research Centre	GEC	East Lane Wembley HA9 7PP, UK
INESC - Instituto de Engenharia de Sistemas e Computadores	INESC	Largo Mompilher 22 Apartado 4433 4007 Porto Codex, P
Telettra Espanola S.A.	TESA	Avenida Cantabria 51 28042 Madrid, E
COSI - Consorzio per l'OSI in Italia	COSI	Via Po 2 00198 Roma, I
IDATE - Institute de l'Audio-visuel et des Telecommunications en Europe	IDATE	Bureaux du Polygone 34000 Montpellier, F

Authors' Addresses

A. Fioretti [1]   Alcatel Alsthom Recherche
Route de Nozay, 91460 Marcoussis, France

A. Bianchi [1]
S. Forcesi [1]   Alcatel Italia, Divisione Alcatel Face
Via Abruzzi 25, 00187 Roma, Italy

D. Capolupo [1]   Alenia S.p.A.
Via Tiburtina km 12,400, 00131 Roma, Italy

A. Fantini [1]   Ericsson Fatme S.p.A.
Via Anagnina 203, 00040 Roma, Italy

E. Neri [1]   ESA - ESTEC
Keplerlaan 1, 2200AG Noordwijk, The Netherlands

F. del Castello   COSI, Consorzio per l'OSI in Italia
Via Po 2, 00198 Roma, Italy

J.P. Bekooij
O. Koning
A. Labrujere
E. Drijver
P. Prinz   Koninklijke PTT Nederland, PTT Research
Sint Paulusstraat 4, 2264 XZ Leidschendam, The Netherlands

G. Veith
H. Schmuck   Alcatel SEL Research Center
Holderaeckerstrasse 35, 7000 Stuttgart 31, Germany

---

1.   At the time this work was done the author's address was:
Alcatel Face Research Center, Via Nicaragua 10, 00040 Pomezia, Italy

G. Cowle
L. Reekie   University of Southampton - Optoelectronics Research Center
                   Southampton SO9 5NH, United Kingdom

B. Hillerich   Daimler-Benz Research Center
                   Wilhelm-Runge-Strasse 11, 7900 Ulm, Germany

A. Almeida
N. Almeida
A. Alves
P. Assis
J. Cabral
E. Carrapatoso
E. Ferreira
M. Ricardo   INESC, Instituto de Engenharia de Sistemas e Computadores
                   Largo Mompilher 22, Apartado 4433, 4007 Porto Codex, Portugal

Springer-Verlag
and the Environment

We at Springer-Verlag firmly believe that an international science publisher has a special obligation to the environment, and our corporate policies consistently reflect this conviction.

We also expect our business partners – paper mills, printers, packaging manufacturers, etc. – to commit themselves to using environmentally friendly materials and production processes.

The paper in this book is made from low- or no-chlorine pulp and is acid free, in conformance with international standards for paper permanency.